D1158922

Christa Sommerer and
Laurent Mignonneau (eds.)

Art@Science

SpringerWienNewYork

Christa Sommerer
Laurent Mignonneau

Advanced Telecommunications Research Laboratories,
Kyoto, Japan

International Academy of Media Arts
and Sciences, Gifu, Japan

Printed in Austria

Typesetting and printing: Adolf Holzhausens Nfg. GesmbH, A-1070 Wien
Graphic design: Yasuhito Nagahara

Printed on acid-free and chlorine-free bleached paper
SPIN: 10566236

With 90 partly coloured Figures

Library of Congress Cataloging-in-Publication Data

Art @ science / Christa Sommerer and Laurent Mignonneau, eds.
p. cm.
Includes bibliographical references and index.
ISBN 3-211-82953-9
1. Art and science. 2. Art and technology. I. Sommerer, Christa, 1964- . II. Mignonneau, Laurent, 1967-
N72.S3A66 1997
701'.1'05--dc21 97-40539
CIP

ISBN 3-211-82953-9 Springer-Verlag Wien New York

Preface

Art @ Science is a collection of articles by media artists and media scientists who over the years have actively forged a collaboration between the arts and sciences. Their interdisciplinary spirit emerged at a time when virtual reality, the Internet, multimedia, and interactivity were yet unknown to the general public.

Through rapid development in digital media technologies and advances in digital telecommunications technologies such as the Internet, computer graphics and interactive media, the human-computer relationship has drastically changed.

Artists working with media technologies in the 70s and 80s often developed their technologies by themselves, giving rise to a whole new genre of artists-engineers, artists-researchers and artists-developers.

On the other hand, traditional research has also changed over the years, allowing itself to increasingly incorporate artistic ideas and approaches to scientific and technological investigations.

What started out in the 70s as an interest by a few interdisciplinary minded individuals has today, with the popularization of digital technology in the 90s, become a major trend in advanced computer graphics, telecommunications and human-computer interaction research.

Parallel to these newly emerging research fields, the arts have also witnessed a major change: a new genre of avant-garde art, so-called "Interactive Computer Art," has emerged over the past 10 years. Using and often developing some of the most advanced technologies, artists in this field are closely connected to questions of human-machine interaction, interactivity and interface technologies.

We both started our careers as artists in modern sculpture and video with additional scientific education in biology and electronics. When we started to use advanced computer technologies as artists, for the first time we recognized the potential of digital technology to make connections among a wide variety of arts and sciences; this allowed us to conduct artistic and scientific research simultaneously. During the last few years we have had the chance to come across many fascinating researchers and artists who have similar personal histories and interests in both the arts and the sciences. All of them are in one way or the other linked to digital computer technologies, such as computer graphics, telecommunication, scientific visualization, media culture, media education, basic research, media art, and interactive media art.

We had the chance to work as artists and scientists at several research centers, such as Advanced Telecommunications Research Laboratories (ATR) in Kyoto, Japan, the National Center for Supercomputing Applications (NCSA) in Champaign-Urbana, USA, and the Institute for New Media in Frankfurt, Germany. What we have witnessed is that a new holistic spirit is beginning to arise, giving hope to an increasingly open-minded and interdisciplinary approach towards the arts and the sciences.

Art @ Science wishes to document this new and fascinating field of artistic and scientific collaboration, and we hope that the reader will find this book to be an important guide along a path which is just now about to open up in front of us.

Acknowledgement
Because this book is a survey, a major regret on our part is that we could not include all of the important artists and scientists working in the field. We do not intend this book to be an encyclopedia of all artist-scientists or all scientist-artists, but hope to document this newly emerging spirit and present some of the most prevalent, current opinions and movements in the field.
We would especially like to thank all the authors for their highly motivated contributions to this book. Furthermore, we would like to thank our publisher, Springer Verlag, and Mr. Siegle, who very enthusiastically supported the project of Art @ Science since the very beginning.
Finally we would like to thank ATR Media Integration & Communications Research Laboratories in Kyoto, Japan for their support in organizing the "ART-Science-ATR" symposium in May 1996, which led to the idea of publishing this book. In addition, we would especially like to thank Professor John Casti for his amiable comments and advice during discussions in Santa Fe and Japan.

Christa Sommerer
Laurent Mignonneau

Contents

Introduction: Art and Science – a Model of a New Dynamic Interrelation

Christa Sommerer and Laurent Mignonneau

I. The Position of Art and Science within Society

When we look back into the history of science we must recognize the resounding influence of Thomas Aquinas, who established a scientific framework during the Middle Ages based on a combination of Aristotelian principles and the precepts of the Christian church. In his system, all questions of understanding and meaning were related to God, ethics and the human soul. Rather than seeking to control or predict nature, his system sought to understand the meaning and significance of the physical world.

During the sixteenth and seventeenth centuries, scientists such as Copernicus, Johannes Kepler, Galileo Galilei, Francis Bacon, and René Descartes started to question this general framework and completely changed the traditional world view of their times, giving rise to the Scientific Revolution. It was Galileo Galilei with his new telescope who established the hypothesis of Copernicus that the world was indeed not the center of the universe. Galileo based his scientific methodology on an empirical approach and a mathematical description of nature. Sir Francis Bacon introduced the concept of scientific experimentation, which profoundly changed science. We can say that after Bacon, the sciences became primarily concerned with the prediction and control of nature. It took the genius of René Descartes to finally establish a whole new system of thought, the Cartesian philosophy, in which *"all science ..."* was to be *"... certain, evident knowledge."* In his quest for the absolute truth, Descartes established a new method of reasoning based upon intuition, analysis and deduction. In Descartes' view of the world, the material universe was a machine that could be described through exact mathematical laws. Descartes' search for "scientific truth" has influenced the general framework of science up to the present day.

During the seventeenth century, the sciences finally won the battle against the church, which until then had been the traditional repository of truth. With the increasing acceptance of science and the Scientific Revolution following the Age of Enlightenment came the belief that knowledge would lead mankind to absolute truth. Science soon became widely accepted as the true path toward security, comfort and social well-being. It has often been observed that in the nineteenth century science became what religion once had been: the only provider of reliable, objective knowledge.

During the later half of the nineteenth century the emphasis on science and especially on natural sciences led to an enormous growth of more specialized subdisciplines. With the triumphant march of science from the seventeenth century through the Industrial Revolution and deep into our lifetime, an increasing separation between the humanities and the sciences arose.

In 1959, C.P. Snow analyzed this dramatic division for the first time.[1]

Snow was a scientist who at the age of 25 became well-known for his discovery of how to artificially synthesize vitamin A. Coming so early to fame, he was regarded as a kind of *Wunderkind.* Unfortunately, he soon realized that his scientific discovery was simply wrong and, under harsh scrutiny from the press, he was soon ruined as a scientist. Later in his career, Snow again caught international attention and finally respect when he formulated a theory for the two cultures of arts and sciences; he pleaded successfully for the reorganization of education and social systems in England and throughout Europe.

At the end of the twentieth century, we are not only confronted with the division of the arts and sciences, or the arts and the humanities, or literature and the sciences (C.P. Snow), but also with the division within the sciences themselves (specialization). Stephan Collini suggests in his introduction to "The Two Cultures," [1] that we can no longer speak about "the two cultures" of the humanities and the sciences, but rather the "two hundred and two cultures."

David Bohm, a distinguished theoretical quantum physicist and one of today's foremost scientific thinkers, remarks that specialization in science has led to an absurd situation, where scientists hardly ever look at fundamental questions, but rather concentrate on details and modifications of their systems.

> The result is to produce an artificial and excessively sharp division between the different problems and to obscure their connections to wider fields.[2]

The German professor Wolf Lepenies notes that science tends to historicize, to defamiliarize, and to view the world in relative terms, but quite remarkably science itself has never become the target of such practices. According to Lepenies, it is time that the *"technique of defamiliarization of the intimate"* be applied to science if we want to overcome blind scientific enthusiasm, which has dominated our view of science since the Age of Enlightenment. He pleads for cognitive flexibility, irony, self-criticism, and the ability to evaluate one's own scientific practice from outside, in order to think about science in relative terms. Production of knowledge may be pursued successfully, but should always be combined with a certain skepticism; furthermore it should become common sense that the enlightenment of science and the consequences of science must become part of science itself.[3]

A similar resistance to outside interpretation can be observed in so-called contemporary art: many artists and art critics consider art to be completely self-referential and one states that, *"Art's only claim is for Art. Art is the definition of Art."*[4] (Joseph Kosuth). Such elitist and socially disconnected thinking has led to the current standstill and imprisoned situation of modern art, which is often more driven by market strategies than by innovative or interdisciplinary thinking.

However, recent movements in the digital arts and especially in the Interactive Computer Arts, which apply new methods, new technologies and new concepts about art and creativity, have given rise to the hope that this antiquated thinking will soon be overcome. "L'art pour l'art" is now a concept that symbolizes the preserve into which art has maneuvered itself within the last century, one which paints art as being disconnected from our recent situation of increased communication, interdisciplinarity and the rise of a global culture. What we are confronted with today is a new type of artist who is

> "... far removed from the ivory tower and close to the hub of radical thought that is shaping the future and informing our destiny." (Roy Ascott)

The romantic movement of the nineteenth century and the philosophy of Immanuel Kant have especially nurtured the notion of the artist as a suffering and antisocial genius, creating out of his own instinct and completely unconcerned with his environment. At the end of the twentieth century, we are now coming to understand that

> "... we need to be more concerned with the role of art as it affects all of us. Art has, of course, always been humanity's highest form of the expression of freedom and imagination. But at the same time, art might have well started, as Jacob Bronowski suggested, as a tool of human survival." (Itsuo Sakane)

It has often been observed that the use of new technologies can cause the loss of an art work's aura, and that the in the age of technical reproducibility the art work has lost it's authenticity. In 1963, Walter Benjamin opened a discourse on the authenticity of the art work, the question of originality and the roles of creator and audience in this new interplay.[5] Benjamin especially examines the influence of photography on the general art discourse and notes that photography was not only recognized as a new tool for artists, but also forced artists to reflect on their art work and its identity in a very novel way. It is also quite significant to the history and the market mechanics of art that it took almost fifty years after its invention in 1840 for photography to be accepted as a real art form.

A similar history is true of film and video. Although film was invented in 1895 and artists almost immediately started using it as an artistic medium, it again took more than thirty years before film was considered to be an artistic medium and for films to be collected by art museums. Video also has led a peripheral existence within the art establishment; although by now almost fifty years old, it is still considered a new media. Video has only recently begun to be accepted as a legitimate art form, especially in the more commercializable form of video installation.

We are now witnesses to a new technological revolution introduced through the invention and rise of digital media, the development of virtual space, and the Internet. Especially in the last five to ten years, an enormous proliferation of new technological and cultural inventions has appeared, completely changing the way we perceive our daily lives, our cultural context, and of course art and science as well. With the invention of the Internet and new telecommunication technologies, people all over the world are now able to communicate and interact, wiping out many cultural and physical boundaries and giving rise to a totally new global culture of virtual and hypothetical space.

Science and technology are influencing daily life in a dramatic way, and in the course of it all, culture and art are being effected as well.

Similar to what Walter Benjamin described as the influence of photography on the art object and on the function of art itself, is the question of how the new digital technologies, CyberSpace and the global culture of the Internet have and will change the role of art and the artist's function. If we are to be conscious about our connection to these important issues of global culture, we will see that this is no longer a question of technology versus culture or art versus technology, but that a synthesis of both will bring about new and unforeseen opportunities for the future.

> Important new insights in science typically have their reflections in other areas of civilization and culture in that they modify or even revise the general world-view or "Zeitgeist." (Gottfried Mayer-Kress)

II. From Happenings and Fluxus to Interactive Computer Installation

The idea that art is connected to technology, science and social issues is of course not new. By 1920, new art forms that contradicted the concept of art as a self referential system and sought dialogue with the audience were already coming into being. The Dadaists provided early examples of such movements: Kurt Schwitters was one of the first to replace the art object by collages of found objects (object trouvée) and early performances (Schwitters Ursonate).

Marcel Duchamp later went a step further and was the first to completely abandon the traditional art object by replacing it with a found object, the so-called ready-made. This was one of the first significant steps, where the artwork got replaced by a reflection the object created in the audience's minds while they were perceiving a common object displayed instead and as an art object. The first artist to stage live action painting before a viewing audience was Georges Mathieu in 1954.[6] By doing so he pioneered the concept of audience participation. Soon after, the composer John Cage first introduced chance procedures into art as a technique for distancing art from the egocentrism characteristic of aesthetic production that had existed since the Renaissance. Cage especially believed that consciousness is not a thing but a process, that art must entail the random, the indeterminate, the chance aspects of nature and culture; he believed that behavioral processes continually inform whether a work of art as an objective state or complete thing, and that

> "the real world ... becomes ... not an object [but] a process."[7]

At the same time artists around the Fluxus movement explored and investigated the definition of art by creating group and individual performances, manifestos, and audience participatory works, forms that allowed direct contact with the audience. Similar to these performances, "Happening," which was invented by Allan Kaprow, proposed simultaneous, polymorphic, multimedia events and actions that gave visual definition to the distinction between art and life, incorporating sight, sound, movement, people, odor, and touch.

With the emergence of video art and early electronic art a new area was born; the new techniques also introduced new concepts about space and time. Nam June Paik is often credited with the development of one of the first interactive works by putting a magnet on a television set, thus transforming the image on the screen through magnetic waves.[8] Bill Etra, Steve Rutt, Stephen Beck as well as Dan Sandin worked on the very early synthesizers, which allowed users to change and modify video images in real-time.[9] Other important pioneers in the field of video and electronic art are Steina and Woody Vasulka who experimented with video and sound since the early 50s.[10]

With advances in computer technology came the discovery of a new artistic medium. Myron Krueger, an early pioneer in the use of computer technologies for interaction, developed his interactive art pieces "Videoplace" in the mid 70s. By the mid 80s individual artists such as Lynn Hershman, Scott Fisher, Ken Feingold, Grahame Weinbren, Jeffrey Shaw and others successively began to use computers for creating interactive art works.[11] The first artist movement to make computer interaction one of its primary goals in the creation and presentation process was a group of artists centered around the Städelschule Institute for New Media in Frankfurt in the early 90s (one might say the "Frankfurt School").

Interaction was defined as audience participation through human-machine and human-computer interaction. A new terminus called "Interactive Computer Installation" was created. Artists like Peter Weibel, Christa Sommerer, Laurent Mignonneau, Ulrike Gabriel, Michael Saup, Agnes Hegedüs, Christian Möller, Akke Wagenaar and others exclusively used computers and interfacing techniques for the creation of audience participatory interactive art works. Other European artists who investigated similar techniques are Edmond Couchot, Monika Fleischmann and Jean-Louis Boissier.

In the United States and Canada, individual artists, such as Michael Naimark, David Rokeby, Bill Seaman, Luc Courchesne, Paul Garrin, George Legrady, Benjamin Britton, Perry Hoberman, Jill Scott and others worked on very similar concepts and techniques. In Japan Toshio Iwai and Masaki Fujihata persued related artistic research. A good overview of the emergence of interactive computer art is given by the book "Pioniere Interaktiver Kunst" by Dr. Söke Dinkla[11] and the article by Dr. Cynthia Goodman in *Art@Science,* Chapter 8.

Interactive Computer Installations can be defined as visual, tactile and/or auditory installations where visitors can interact with images, sound, textures and artificial environments that are in most cases created through digital technologies. They all use real-time interaction as a significant characteristic: this real-time aspect allows the visitors to these installations to actively participate in the process of image or sound creation and to modify parameters or even the whole work of art itself. These works use different interactive technologies such as sensors and interface devices to allow real-time interaction. In many cases, the artists not only create the art work but also invent and develop sensors and interfaces themselves, closely involving them with research and design. By providing an audience modulated framework, these interactive art works become variable, modifiable and personally sensitized to the audience's interaction. In chapter four of this book, artists such as Professor Jeffrey Shaw and Professor Peter Weibel, the artist-researcher Michael Naimark, the researcher-artist Monika Fleischmann and the editors themselves describe some of these interactive computer installations.

III. Interactivity and the Interconnected Web

The concept of the interaction, interrelationship and interconnection of entities began to appear in science when in 1950 Relativity and Quantum theories were developed by Albert Einstein, Erwin Schrödinger, Werner Heisenberg, Max Planck, and Niels Bohr.

Since the establishment of Relativity theory and Quantum physics, we have come to understand that the observer, the human mind, is an integral part of the observed structure, and therefore affects it: *"The mind of the observer (scientist) will be thus reflected in the result."* [12] Quantum theory has shown that subatomic particles are probability patterns and interconnections in an inseparable cosmic web that includes the human observer and eventually even his or her own consciousness.

Today, the activity of the cosmic web is coming to be understood as the very essence of being, as one indivisible cosmic whole; the single individual is no longer considered essential, but instead the interaction and the transformations of single entities has become the focus of research and consideration.

Modern theories of physics are giving rise to a new world view, that is no longer the objective, exclusive Cartesian-Newtonian worldview dividing mind and matter. For example, David Bohm's theory of quantum physics treats the totality of existence, including matter and consciousness, as an unbroken whole.[13] Chew's S-Matrix Theory contends that human consciousness may very well be an essential part of the universe. Both of these theories agree that patterns of mind and patterns of matter are reflections of one another and that every particle consists of all other particles. These particles are not separate entities but components of patterns that involve one another. This concept of involvement is described very beautifully by the biologist and anthropologist Gregory Bateson, who has examined the influence of this thinking on our social structures. Bateson has proposed a hypothesis of how the mind within an organism or even a larger social structure can develop through an aggregate of interacting parts.[14] In the wake of the Cartesian division of mind and matter, we can now finally witness the reintroduction of mind as an interrelated and essential component of matter in the research process.

Observation and analysis of recent developments in interactive media art and new digital media art technologies reveal a similar and complimentary philosophy at work in the most recent avant-garde art movements. In interactive computer installations, as well as web art and Internet art, matter is now being reintroduced into the creative and artistic research process. The emergence of the global brain (as described in the contribution by Gottfried Mayer-Kress) is just one example of a hyper-structured convergence of mind and matter.

What we want to suggest in this book is that in this new worldview containing dynamic webs of interrelations and interactions, human consciousness is an important link between mind and matter.

Similarly, we suggest that art and science should no longer be considered separate and contrary disciplines, but instead complementary to each other, where patterns of mind (art) and patterns of matter (science) are reflections of one another that are dynamically interrelated through the human consciousness, changing their states (just like electrons and neutrons) from mind to matter and vice versa from matter to mind. We consider both of them part of a holistic, intrinsically dynamic and self consistent universe.

IV. A New Holistic Thinking

At the end of the twentieth century and the dawn of the new millennium, we now observe a new spirit of interdisciplinary and holistic thinking that will shape the future in the next millennium.

> And it is about time that we free ourselves from the causal, analytical, reductive, linear time axis elemental explanations which have ruled our 20th century world view. (Toshiharu Itoh)

This spirit of Aufbruch, or breaking up, is especially characteristic for the new computer technologies and foremost for the interactive computer arts and digital arts.

> Interactive computer graphics has become a shared language in many fields of research, and as a consequence a great diversity of information coexists that can be correlated in the digital environment. This is a unique situation historically and culturally, one which artists and scientists can take advantage of to forge a new discourse. (Jeffrey Shaw)

Although still in their infancy and specialized in fields such as telecommunications and computer based technologies, it is becoming more and more clear that interactive technologies are not just another interdisciplinary approach. Interactive technologies together with interactive art movements will come to have a larger cultural impact, because they finally forge a true break with the traditional Cartesian division of mind and matter and the segregation of art and science. In the emerging fields of Interactive Computer Art and graphical computer science, art and science increasingly influence and fertilize each other. Results can be applied not only to engineering, research and technology, but are significant for art and the discourse of art as well.

> Such is the complexity in technoetic art of the relationships between mind, behavior, environment and technology that research is now at the basis of many artists' practice. (Roy Ascott)

What especially characterizes this interdisciplinary holistic spirit of the interactive arts is its international scope and its wide acceptance among the general public. The artists working in interactive media are not only working and presenting to the art elite, but regard the general public as their primary audience.

We now understand that an overemphasis on the Cartesian method leads to reductionism in science, and arguably, also to reductionism in the arts. The critical and open-minded scientist or artist nowadays should realize that cultural progress can only take place through discourse.

This influence is mutual. Not only are artists starting to collaborate with scientists, but also scientists are more open to the new possibilities of this interchange.

> Science is a child of compassion. To check whether the world is consistent is a monk-like endeavor for many generations. It brings all kinds of fruits. The steel-fibers of analytic geometry [..] on the one hand enlarges the arsenals of the military to unprecedented proportions; on the other hand, medicine - the plumbing of the body as a deterministic machine – enables the survival in health of billions; third, the computer, with the vertical exteriority that it conveys relative to artificial universes, enables a new understanding of the world by analogy, and the horizontal exteriority of the Internet enables the first new evolution (and the first democracy). (Otto E. Rossler)

Forward looking scientists who are tired of the excessive specialization and artificial dissection of science into more and more subdisciplines that lose the connection to the general socio-cultural background, have discovered a new view of science itself.

> The history of the past two decades has shown a dramatic change in the direction of scientific evolution back to the more direct experience of our daily life. This change of direction was influenced to a large extent by our enormous progress in both computer and communication technologies. At the same time our daily life experiences have changed in a similar dramatic way by exploring another new frontier: CyberSpace. (Gottfried Mayer-Kress)

And pioneer Myron Krueger points out the following:

> The difference between artists and scientists is that artists believe in their intuition, scientists don't. And engineers are paid to trust in other people's intuition. However progressive science is made by scientists who do believe in their intuition.[15]

V. Areas of Interaction between Art and Science

For this book *Art @ Science*, we have invited scientists and artists who are well known and interested in a global, holistic and interdisciplinary approach to give their views. These scientists and artists have been actively establishing and creating this new mutual influence and interrelation. It is not by chance that most of them are connected in one form or another to new computer technologies and many of them are pioneers in their fields.

It is characteristic of these new media technologies that they are in themselves already interdisciplinary, so the separation of chapters may seem somewhat artificial; in fact many individuals would fit into one or more categories. Nevertheless, to provide a structured overview, we have divided the book into areas where fertile art and science interaction is taking place. The resulting chapters are as follows:

1.Telecommunications; 2. Scientific Visualization; 3. Artificial Life; 4. Artists as Researchers; 5. Chaos and Complex Systems; 6. Public Spaces; 7. Education of Art & Science; 8. Art & Science in Historical and Cultural Context.

Telecommunications (Chapter 1) is an area that has seen a vast technological revolution over the years and will continue to have a deep impact on our society in the new millennium. With the emergence and the establishment of the Internet and virtual telecommunication technologies as a whole new field of application, new types of research, commerce and an on-line culture has emerged. Virtual communities, virtual companies, virtual information networks, and virtual cultures are proliferating.

In the field of virtual space and in its connection to the latest telecommunication technologies, the links between science, research, art, and culture have strengthened over the last 5 years. Large telecommunication companies such as Nippon Telegraph and Telephone (NTT), Deutsche Telekom and Telefonica Spain sponsor cultural centers such as the Intercommunication Center in Tokyo Japan (NTT-ICC), the Guggenheim Dependence of the ZKM in New York (German Telecom), and the prolific CyberFestival in Madrid Spain (Telefonica Spain). They also work towards public understanding and acceptance of new media technologies. Furthermore, these telecommunication research centers have now begun to integrate artists into the research process for developing new telecommunication technologies. One example is the Advanced Telecommunications Research Laboratories (ATR) in Kyoto, Japan, where artists work closely with scientists to develop new technologies and concepts of verbal and non-verbal communication in virtual space. Dr. Ryohei Nakatsu, president of the ATR Media Integration & Communications Research Laboratories, facilitates a new type of collaboration between scientists and artists based on the following consideration.

> As the research has advanced, however, it has begun to be recognized that behind the scene of logical communications, a more basic mechanism of communication, based on human emotion and kansei (sensitivity), plays a very important role for mutual understanding between humans. This basic form of communication can be called interaction. It is indispensable to study the mechanism of interaction and to develop technologies that can realize highly human like communication, by integrating communication and interaction technologies, as well as interactive arts.[16]

He believes that in the collaboration of art and science,

> ... there is a good chance that art and technology can work together to realize highly sophisticated communication methods and art pieces; in other words, communication methods that can overcome the cultural and language gaps among people.[16]

An overview of Telematic Art, or Communication Aesthetics, is provided in an article by Maria Grazia Mattei, an Italian art historian and critique. She examined the rise and evolution of telecommunication art, which originated in the late 70s. This movement was tightly linked to the commercialization of major advances in telecommunication devices such as telephones, FAX machines, satellites, and computers. Artists and scientists in this area have closely collaborated to explore creative possibilities of the new media. In her article, Mattei introduces the most influential pioneer thinkers in Telematic Art.

Philippe Quéau, director of the Information and Informatics Division at UNESCO in Paris and former director of the annual Imagina Festival in Monte Carlo (one of Europe's most important public events in this area of telecommunications, computer graphics and commerce), is looking into the deeper meaning of telepresence, televirtuality and the questions of presence and representation. As we are now able to create virtual communities through new communication technologies such as virtual simulation techniques, virtual reality, CyberSpace, the use of avatars, and augmented reality, we will also need to question what it means to be really present or virtually present. In his essay, Philippe Quéau looks into these deep philosophical questions and draws connections to Husserl's phenomenology.

Another field where a constructive collaboration between art and science has taken place over the past decade is *Scientific Visualization (Chapter 2)*. Although not primarily artistic and originally regarded more as an applied art form for the demonstration of scientific research results, it has since freed itself to become a hot new area where artist-engineers and artists-scientists work with scientists. Professor Dan Sandin of the Electronic Visualization Laboratory at the University of Illinois at Chicago is a renowned pioneer and inventor of the CAVE Virtual environment system. Professor Sandin, together with Professor Tom DeFanti, has especially helped to establish this new and extremely productive area of collaboration. Professor Sandin's work is described by Dr. Cynthia Goodman in her article in Chapter 8, "Art & Science in Historical and Cultural Context."

Professor Donna J. Cox, a computer graphics artist and one of the most important figures in scientific visualization, has been working on collaborating art and science over the past decade. Her scientific visualization, "Cosmic Voyage," which represents colliding galaxies, was solely rendered on a high-end graphics work station and super computers. "Cosmic Voyage" was realized for the IMAX cinema and the Smithsonian Air & Space Museum in Washington DC. This work has won Professor Cox and her team the renowned Academy Award of best scientific visualization.

> The extensive use of high technology requires collaboration. Scientific visualization involves a variety of artistic, technical, and scientific expertise. The "Cosmic Voyage" is an excellent example of artists working with scientists to produce graphics that are fantastic and awesome. This type of production would be impossible without global collaboration.

Professor Przemyslaw Prusinkiewicz from the Department of Computer Science at the University of Calgary has been conducting research in computer graphics since the late 70s. In 1985, he originated a method for visualizing the structure and development of plants based on the famous L-systems (Lindenmayer Systems), a mathematical model of development. He is coauthor of the bestseller, The Algorithmic Beauty of Plants in Springer-Verlag, 1990,[17] and the Lindenmayer Systems, Fractals and Plants in Springer-Verlag 1989. Professor Prusinkiewicz's scientific models for the creation and development of artificial computer generated plants and natural structures have influenced many scientists in the various fields of computer graphics and his books have achieved the status of classics in this field of computer modeling. In his article, Professor Prusinkiewicz looks at the question of what defines a good scientific model and where to draw the line between abstraction and strict realism.

> ... modeling is an interdisciplinary activity which combines a knowledge of the modeled phenomena, the computer science methodology for creating the simulations, and the craft of visualizing the models for validation purposes. In its highest form, this craft becomes an art, fulfilling the role phrased by Pablo Picasso: Art is the lie that helps us see the truth.

Professor Demetri Terzopoulos of the University of Toronto, Canada, is a highly distinguished researcher in the field of animation, scientific visualization and artificial life. In his Computer Science and Electrical and Computer Engineering Laboratories, he is leading a Visual Modeling Group, which is developing novel ways to model realistic movements of virtual humans and virtual life, and it is described in this book. Especially noteworthy is his work on artificial fish. The behavior of virtual and artificial life has gained him international recognition, making him an important figure and a member of many international editorial boards of journals including the *Journal of Visualization and Computer Animation.* Professor Terzopoulos is also an organizational member of the annual Siggraph Conference, the world's most important computer graphics conference in the United States.[18]

Artificial Life (Chapter 3) is a field of research that was for the first time described by Christopher Langton in 1987 at the first Artificial Life workshop held at the Los Alamos National Laboratories in the United States.

> Artificial Life is a field of study devoted to understanding life by attempting to abstract the fundamental dynamical principles underlying biological phenomena, and recreating these dynamics in other physical media (such as computers) making them accessible to new kinds of experimental manipulation and testing. ... In addition to providing new ways to study the biological phenomena associated with life on Earth, "life-as-we-know-it", Artificial Life allows us to extend our studies to the larger domain of the "bio-logic" of possible life, "life-as-it-could-be" whatever it might be made of and whatever it might be found in the universe.[19]

One of the pioneers of Artificial Life (A-Life) and the founder of the "Tierra" system is the biologist Dr. Thomas S. Ray. Tierra is the first computer generated universe where virtual life is able to propagate, reproducing and evolving in the form of an executable computer code. Dr. Ray has recently created an on-line version of Tierra, and he is now studying the impact of the networked, shared living space of these organisms that can migrate from continent to continent and so create a biodiversity of virtual life forms. In his article, Dr. Ray describes his research on Tierra and also introduces the important work of A-Life artist Karl Sims. Karl Sims is one of the first artists to be inspired by the possibilities of Artificial Life, and he has created interactive computer graphics works such as "Genetic Images" and the graphics animation "Panspermia."

The first artist to creatively explore Artificial Life was zoosystematician Professor Louis Bec. In 1972, he founded the Scientific Institute for Paranatural Research in France, where he continues to conduct artistic and scientific research on Artificial Life and its relationship to the living. In his article "Artificial Life under Tension," he examines the "tension existing between life defined as an intrinsic property of matter and life redefined as a technological simulation device." To him, "this tension describes a distinctive trajectory in the overall relationship between the arts and the sciences."

Artificial Life has since inspired many artists. Professor Machiko Kusahara from the Tokyo Polytechnics University in Japan is an important writer and curator in A-Life art. In her article, she introduces various artistic approaches and artists in this field with an emphasis on the Japanese recent situation. The editors of *Art @ Science* have, for example, worked with Dr. Thomas S. Ray to create their interactive computer installation "A-Volve;" virtual interactive creatures live in a water filled pool where they reproduce, mate, compete, and evolve. This work was the first interactive work where visitors could actually create Artificial Life by themselves, watch it evolve and also interact with it through touch.

Artists as Researchers (Chapter 4) introduces some of the most prevalent artists in the new field of interactive media art. This chapter provides a selection of artists who specifically work as artists at research centers. For example, Michael Naimark is one of the pioneer artists in interactive media and his work has been shown world wide. For several years Naimark has worked with the Interval Research Corporation of Palo Alto, California, where he is pursuing artistic and scientific research on virtual reality and new interface design. Interval Research Corporation was founded by Paul Allen, co-founder of Microsoft Corporation, and defines itself as a center seeking ...

> ... to define the concerns, map out the concepts and create the technology that will be important in the future. With its long-term resources, Interval pursues basic innovations in a number of pre-competitive technologies and seeks to foster industries around them – sparking opportunity for entrepreneurs and highlighting new ways of researching technology. To bring a fresh perspective, Interval has gathered a broad range of people to make up its research staff, including filmmakers, clothes designers, musicians, cognitive psychologists, artists, computer scientists, journalists and software developers.[20]

Artist Monika Fleischmann and Professor Wolfgang Strauss are working at the GMD Gesellschaft fuer Mathematische Datenverarbeitung in Bonn, Germany. GMD is Germany's National Research Center for Information Technology, where they conduct research in informatics, communication and media.[21] Fleischmann's artistic research in the Visualization and Media Systems Design Laboratory is concentrated around the design of novel visualization and media art systems. In her article, she describes various collaborative projects between artists and scientists such as "Responsive Work Bench," "Home of the Brain" and "Liquid Views."

Professor Christa Sommerer and Laurent Mignonneau work as artists and researchers at the ATR Media Integration and Communications Research Laboratories in Kyoto, Japan,[22] and teach interactive media art at the IAMAS Academy of Media Arts and Sciences in Gifu, Japan[27]. Sommerer and Mignonneau are focusing on the development of interactive computer installations that allow personalized non-deterministic and multi-layered interaction and real-time access to virtual space. To provide individual feedback and personal expression to the visitors of their installations, they have developed unique and non-deterministic natural interfaces, such as living plants in their work "Interactive Plant Growing" or water in their interactive computer installation "A-Volve". In 1992 they developed the concept of "Natural Interface," which has since been widely exploited in research as well as art. As described above, their research also includes Artificial Life and Complexity as well as new telecommunication technologies.

Professor Jeffrey Shaw has been working in Interactive Media and Virtuality as art forms since the 60's. In his many art installations such as "Legible City," and the "Virtual Museum," he allows the spectator real-time access to virtual space. At present, Shaw is director of the Institut für Bildmedien, the Institute for Visual Media, at the Zentrum für Kunst und Medientechnologie (ZKM) in Karlsruhe, Germany.[23] He leads a unique research and production facility where artists and scientists are working together and develop profound artistic applications for new media technologies. In his article for *Art @ Science,* Professor Shaw describes the impact these new technologies can have on art and especially on interactive media art, and he suggests possibilities for fruitful collaboration.

Professor Peter Weibel is the founding director of the Städelschule Institute for New Media in Frankfurt Germany and Professor for Media Art at the Academy of Applied Arts in Vienna. As exhibition curator, Prof. Weibel for many years has been the artistic director of the renowned "Ars Electronica" festival in Linz, Austria. He has curated many important exhibitions on art and media, including the Venice Biennale 95. At the Institute for New Media, Professor Weibel was instrumental for establishing Interactive Computer Art as a new art form and successively he has educated many of the now well-known media artists in this field. In his article "The unreasonable effectiveness of the methodological convergence of art and science," he critically and meticulously examines the convergence of art and science from the past until the postmodern and multimedia ages, starting with the famous paragon dispute in the arts, to the question of whether art can be considered a method (as do the sciences), to finally introduce a common ground between arts and sciences, a social construction.

Besides telecommunications, scientific visualization, Artificial Life and computer graphics, chaos and complex systems is another research area where significant cross-influence between art and science has taken place over the years.

Chaos and Complex Systems (Chapter 5) presents writings by chaos scientists and researchers in complex systems science and nonlinear dynamics.

Dr. Gottfried Mayer-Kress from the College of Health and Human Development at Penn State University in the United States describes in his article how the

> ... advent of non-linear dynamics and chaos theory suddenly changed the whole scene: One discovered basic science frontiers out there that are not found in the domain of the very large or very small but in the new dimension of non-linearity. We became aware that science did neither understand the movement of a few planets (the question of the future of our solar system was not known because of chaotic predictability limitations) nor the behavior of a child's swing or a dripping faucet.

The discovery of the non-simple (complex, complicated, erratic, unpredictable) phenomena in our lives has also influenced the way scientists have viewed the world. In his excellent article, Dr. Gottfried Mayer-Kress introduces how chaos science and complex systems science and the research of non-linear dynamics have influenced the field of art, and he describes how artists have been starting to use this new knowledge as a source of inspiration in the forms of the Sound of Chaos or Brain Waves or Fractals. He also introduces the concept of the "Global Brains," which emerges from the new possibilities of CyberSpace and the structures of the globally connected Internet.

> Through the high degree of global connectivity that is becoming available for the first time in human history it now becomes likely that coherent structures or order parameters will emerge with shared properties and a coherence that has not been encountered before.

The famous chaos scientist and inventor of the Rossler attractor, Professor Otto E. Rossler from the University of Tübingen in Germany, describes in his essay how Descartes' thinking was influenced by his artistic and compassionate spirit.

> If art consists of making enlightenment visible, he [Descartes] was an artist. [..] Descartes did not know whether this logical dream of his would last ten minutes or ten days or a year. But he knew that as long as this hope was unfalsified, he could not rule out being in the role of a superprogrammer himself.

Public Spaces (Chapter 6) introduces new museums, centers and other public spaces that are now being established all over the world to present the interrelation between electronic media, latest telecommunication developments and new interactive media art. One of three most prolific centers is the InterCommunication Center (ICC) in Tokyo Japan,[24] which is mainly sponsored by the world's largest telecommunication company, Nippon Telegraph and Telephone (NTT). Professor Toshiharu Itoh, one of the founders and conceptual leaders of the ICC, explains the educational and cultural importance of such centers for an information society in Japan and internationally.

> The rapid progress information-oriented societies have altered not only the face

of industry and economy, but brought about great changes in the society and culture. [..] The ICC is a center for taking the leading edge electronic communication technologies core to NTT operations, and forming a broad-spectrum experimental link for how they will come to affect our society and culture.

Germany's pioneering center for media art and media culture is the Zentrum für Kunst und Medientechnologie (ZKM) in Karlsruhe[23], and could be considered the European equivalent of the ICC Center in Asia. Professor Hans-Peter Schwarz, the director of the Media Museum of the ZKM, explains the importance of this new type of media art museum. As a cultural historian, he has initiated discussions about how new digital and interactive media art challenge our perceptions and the question of the art work's authenticity, and its embodiment in a general art context.

The third of the three large media art centers that have recently been established is the AEC Ars Electronica Center in Linz, Austria.[25] The annual media art conference, Ars Electronica, and the Prix Ars Electronica were essential for the development, support and presentation of media art and media sciences over the past 20 years.

The Exploratorium[26], which was founded in 1969 by Frank Oppenheimer in San Francisco, is one of the most active centers in the United States. There cultural and scientific discourse are activated through exhibitions, public presentations and an educational program in the sciences and arts. Peter Richards, the director of the exhibition program, introduces the concepts of the Exploratorium and its socio-cultural impact and focus.

It has been noted that a common ground between artists and scientists is their process of exploration, experimentation, and testing of intuitions. These kinds of activities precede, and are germane to all scientific and artistic creation, and are inherent in children. This playful attitude has become a common value throughout the Exploratorium, a museum of science, art, and human perception, providing a unifying thread for creating meaningful and enjoyable experiences for the public.

Education of Art & Science (Chapter 7) deals with the issue of how education of art and science has become more and more important in our daily lives and the public understanding and training in these newly emerging fields of media art and media science. When we look, for example, at the situation in academia (and this is more true for Europe than for the United States or Japan) we see that institutes and academies only sporadically offer classes about computer art and interactive media. This situation is quite alarming as the art establishment already seems to have lost today's cultural pulse and fallen behind the new movements of social and global culture. A knowledge gap of at least ten years of activity in electronic media art and media science must be filled to prevent art academia from completely losing its connection to the awakening electronic and digital media culture.

In the education of today's art student, catching up and acquiring knowledge will play a significant role if he or she is going to be part of the global culture. The importance of such educational upgrading, introduced in 1959 by C.P. Snow, is now being recognized internationally; in pioneer countries of digital media such as Japan, the importance of an overall art and science education is not underestimated. Professor Itsuo Sakane, the pioneer and father of the art-science movement in Japan has for example recently founded a new International Academy of Media Arts and Sciences (IAMAS) in Gifu, Japan,[27] where education in media art and media science is shaping the future of Japan's art and media elite.

The Institute for New Media in Frankfurt Germany[28] was originally led by Professor Peter Weibel and is now under the direction of Dr. Michael Klein. Dr. Klein is by education a chaos scientist and former student of Professor Rossler. In his excellent article, he describes how he moved from an interest in chaos science toward the interactive arts, and how his institute is constructively working for the education of media artists and media scientists. Especially in the case of the Institute for New Media, one can observe how little support from the education system these kinds of new and progressive institutes actually receive. Indeed, they are often forced to operate as private schools that have to seek their own funding. If we are to catch up with today's rapid developments in global culture, it will become increasingly important that such institutes be integrated into and supported by the publicly available education sector.

In *Art & Science in Historical and Cultural Context (Chapter 8)*, media art historian, curator and author of the book Digital Visions: Computers and Art [29], Dr. Cynthia Goodman presents the development of new digital and interactive art from its beginnings in the 60s until today by introducing the most important artists and movements in this area. Dr. Goodman has recently edited a CD Rom compendium about this new interactive media art, the "InfoArt" CD Rom and catalogue, which gives a very good overview and introduction to this field.[30]

Professor Erkki Huhtamo, writer, curator and Professor at the University of Lapland in Finland, investigates specifically the archeology of media art and in particular presents and analyses the archeology of the small screen, which to him symbolizes a window into the virtual world, similar to the mirror in Jean Cocteau's film Orphée (1950). Huhtamo's Media Archaeology research is focusing on the larger cultural context within which new technological developments are embedded.

> Technology is not conceived in terms of a linear, progressing series of gadgets that make their predecessors obsolete. [..] The important thing is to understand the role of technology (in this case the technology for showing and/or transmitting images by means of specifically designed surfaces) as one of the ingredients in cultural processes.

In the same chapter, the writer, curator and artist Professor Roy Ascott analyses the technoetic dimension of art and draws a general connection between art, technology and its cultural context in today's global network. Professor Ascott also recently founded the Centre for Advanced Inquiry in the Interactive Arts (CAiiA) [31] at the University of Wales, in Newport, UK, which pursues advanced research in the technoetic domain as a virtual community.

References

1 Snow CP (1993) The two cultures. Cambridge University Press, Cambridge, MA
2 Bohm D (1987) Science, order and creativity. Routledge
3 Lepenies W (1986) Vom Enthusiasmus zur Skepsis in Technik und Wissenschaft. In Computerwoche. Nr. 45
4 Stiles K, Selz P (1996) Contemporary art – a source book of artists' writings. University of California Press, pp. 847
5 Benjamin W (1977) Das Kunstwerk im Zeitalter seiner technischen Reproduzierbarkeit. Suhrkamp Verlag
6 Stiles K, Selz P (1996) Contemporary art – a source book of artists' writings. University of California Press, pp. 698
7 Cage J, Daniels C (1981) For the birds: John Cage in Conversation with Daniel Charles. Boyars, Boston and London, p 80
8 Hong-Hee K, Goodman C (1995) InfoArt – 95 Kwangju Biennale. Kwangju Biennale Foundation, pp 102–108
9 Dunn D, Weibel P (1992) Eigenwelt der Apparate-Welt, Pioneers of Electronic Art. Linz: Ars Electronica, pp 91–160
10 Vasulka S, Woody (1996) Machine media. San Francisco Museum of Modern Art
11 Dinkla S (1997) Pioniere Interaktiver Kunst. Edition ZKM. Cantz Verlag
12 Capra F (1982) The turning point. Bentam Books, pp 75–97
13 Bohm D (1980) Wholeness and the implicate order. Routledge
14 Bateson G (1979) Mind and nature. Bantam Books
15 Krueger M (1997) In a personal interview with editors on March 10, 1997 in Gifu, Japan. Further readings: Krueger M (1991) Artificial Reality II. Addison-Wesley, Reading, MA
16 Nakatsu R (1996) Opening statement for "ART-Science-ATR" conference on May 13–14, 1996. http://www.mic.atr.co.jp/~christa/Alink2.html
17 Prusinkiewicz P, Lindenmayer A (1990) The algorithmic beauty of plants. Springer, Berlin Heidelberg New York Tokyo
18 Siggraph Home Page: http://www.siggraph.org
19 Langton C et al. (1992) Artificial Life II. Addison-Wesley, pp. xiv
20 Interval Home page: http://web.interval.com/about
21 GMD Home Page: http://www.gmd.de
22 ATR Home Page: http://www.atr.co.jp
23 ZKM Home Page: http://www.zkm.de
24 ICC-NTT Home Page: http://www.ntticc.or.jp
25 AEC Home Page: http://www.aec.at
26 Exploratorium Home Page: http://www.exploratorium.edu
27 IAMAS Home Page: http://www.iamas.ac.jp
28 INM Home Page: http://www.inm.de
29 Goodman C (1987) Digital visions: Computers and art. Harry N. Abrams, Inc.
30 Goodman C (1996) InfoArt CD Rom. Rutt Video Interactive, http://www.rvi.com
31 CAiiA Home Page: http://caiiamind.nsad.newport.ac.uk

1. Telecommunications

Virtual Communities: The Art of Presence

Philippe Quéau

Different kinds of virtual communities are rapidly developing. From MUDs to LambdaMOO, from SIMNET to ExploreNet, from Habitat to WorldChat, V-Chat, CyberTown, Moondo, ... there are hundreds of variations of the concept of "being virtually together"... Other communities will not wait long before taking advantage of the future developments of technology. Virtual networks of artists, virtual companies, virtual traders, virtual gamblers, virtual information networks are proliferating. The development of powerful simulation technologies including realistic televirtuality applications and efficient groupware techniques will only facilitate the spreading of this long term trend.

At the Televirtuality Group at INA we are working on real time analysis of facial characteristic and real time realistic rendering of facial expressions.[1] We think that the main interest of the technique thus developed is to allow virtual communities to get a better sense of "presence" through the virtual representations of 3D faces with a realism faithful to the original (Fig. 1).

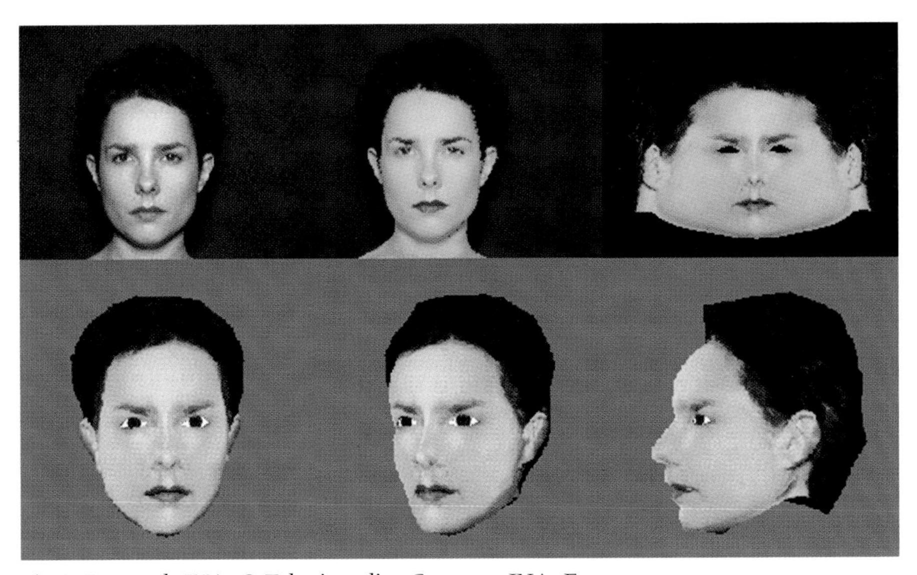

Fig.1. Research INA. © Televirtuality Group at INA, France

I would like now to focus on a rather philosophical question: we will be more and more confronted to various forms of "virtual presence", with various impacts on "real reality".

Telepresence, augmented reality, televirtuality are examples of different degrees of presence, using different represention and different ways of mixing "real presence" and "virtual representation".

It will be increasingly difficult to recognize the level of reality and of virtuality which we will be immerged in. It is not merely a problem of mistaking realistically looking "avatars" for "real humans", like in a *Blade Runner* world. It is more of an epistemological problem. As virtual instances of reality are proliferating, the frontier between reality and virtuality will blur. Before answering to the question: "what is virtual?", we will have to answer firstly to another question: "what is real?". Is a quark "real"? What is the "real" value of the yen? Is an economic bubble "real"? A real-virtual dialectic is apparently developing to the extent of destroying many certitudes.

Another aspect of the problem deserves a special attention: the more virtual communities develop, the more real ghettos and real social fractures are also developing: why is that so?

Virtual communities propose a new way to represent "presence", a new way to be present to other people, to be "together" in a virtual manner. In their own times, painting, photography, cinema used to introduce also such innovative ways to represent "presence". Thus technology is somewhat related to the way we are "present", and to the way we perceive the essence of being present. The essence of representation is related to the representation of presence. A new representation modifies our understanding of presence. Any presence modifies our understanding of its representation. The relationship between art and science is comparable to the relationship between presence and representation. The art of representation is related to the science of presence.

What is presence? What is representation? Those two terms are contradictory. But with televirtuality and virtual communities, we "meet" avatars which are "present" though they are "representations". Our categories do fall apart.

Virtual avatars are submitted to delocalization, dematerialization, desintermediation. They are not really "there", yet they do transmit a sense of presence, of being together, out there in the cyberspace.

The question we'd like to address is the following: to what extent does art help us to represent (in a convincing way) presence? How does art help us to be present to ourselves? How does art help us to be present to other people?

The words "presence" and "representation" are strictly opposed, at least in French or in English. It's embedded in the language itself. In Japanese, one could translate presence by: sonzai, shusseki, menzen, fûsai.

Sonzai is more like "existence" like in:

Are you sure of the presence of ghosts? (Kimi-wa yûrei no sonzai o-shinjiru-ka?)

Shusseki is like "attendance" like in:

Your presence is requested. (Go-shusseki o-negaimasu)

Menzen is like "someone's face" like in:

Don't scold your sons in the presence of others. (Tanin-no-menzen-de musuko-tachi o-shikattewa-ikenai)

Fûsai is like appearance, air like in:

He has a poor presence. (Kare-wa fûsai ga agara-nai.)

One could translate representation in Japanese by: daihyô, shôchô, hyôgen, egaki.

Daihyô is like exemplary, typical.

Shôchô is like a symbol, an emblem.

Hyôgen is like an expression, a manifestation.

Egaki is like a portrait, a picture.

Thus, in Japanese too, presence and representation are somewhat distant concepts.

What we'd like to emphasize is that the new digital image technologies make it more and more difficult to draw a dictinct line between those two concepts.

Edmund Husserl introduced the concept of *epochè* as a central concept of phenomenology.[2]

Epochè could be described as a "suspension of belief". Confronted to the world, the phenomenologist would sort of stop "believing" in it, just to see what would happen from this radical doubt, suspecting that from this very doubt new insights might emerge.

We'd like to compare any representation to an *epochè*. A representation is a suspension of belief: when we think that a "presence" is not really possible, we stop believing in a real "presence" and we substitute it by an ersatz that we call a representation. Representation is a presence in italics, a "presence" in parenthesis.

Thus we are led to think that there might be different "levels" of presence. Any representation is a kind of weakened presence, a "presence" which lacks of presence. A representation is not a re-presentation, that is to say a "renewed" presence. Representation introduces a distance.

In French and in English, the word "present" has a double meaning: it means "now" and it means "gift". The feeling of being "now" is a gift. The ultimate gift is the possession of "now-ness". You give more of yourself when you are present than when you are represented.

A classical way to mix presence and representation was the theater use of masks. In Latin the word "persona" had three meanings: mask, character, person. These contradictory meanings are symptoms of a fundamental problem: how can we represent what has no representation, what is irrepresentable? How can we represent the divinity? The mask is a sort of an answer to this challenge. The mask hides the presence of what is essentially without any representation. In the Greek mythology, the Gorgô had a face which was a mask. The face-mask of Gorgô was a sex: the sex was just another mask. This provoked the laughter of the gods, exactly like in the Japanese mythology (cf the story of Amaterasu-ô-mi-kami).

When one wears a mask, one does cease to be what we really are. We become the image of the mask for others.When one wears a virtual reality helmet, what do we become for others?

Now we can ask ourselves again: how should we try to represent presence? We know it is an oxymoron. Maybe "art" could be helpful to overcome it.

Do we mean presence as a "now", as a "present", or do we mean it as an "existence", an "attendance", a "face", an "appearance"? Do we mean it as a mixture of these meanings?

We would like to introduce here two examples of "presence", or more precisely two examples of a representation of presence.

The first one is drawn from the work of the artist-photographer Emmanuel Carlier. He works on presence as a "now". He is an artist of the presence caught as a still image. Using multiple cameras, he freezes time and travels at will "in" the virtual space of the "present". This gives a new depth, a new thickness to the presence of the bodies. The body becomes then a space that can be explored like a virtual land, an imaginary planet.

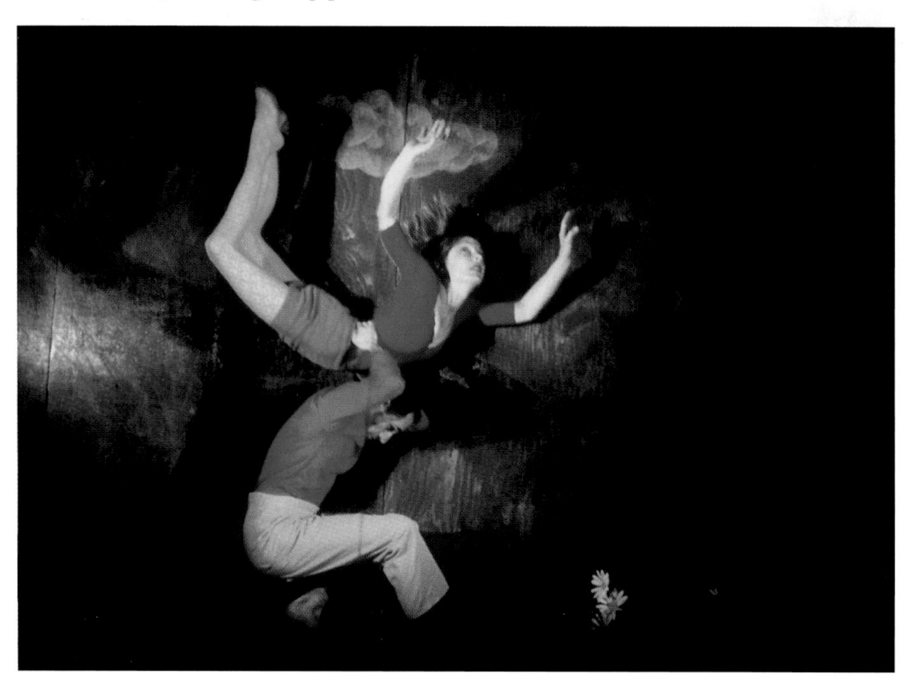

Fig. 2. Gravite Zero. © Kitsuo Dubois, France

Another example comes from the artist-dancer Kitsou Dubois (Figs. 2, 3). When Carlier just showed us bodies as explorable spaces, Dubois wants to show us bodies exploring different spaces.

The word "space" should be taken literally. She studies what becomes her art of dance, her ability to move her own body in a zero gravity environment.

With Carlier presence is *a space* to explore.

With Dubois presence is *in space*;[3] the space reveals another level of presence of the body, another awakening of the body to its own structure.

With Carlier, presence is a time that can be unrolled, explored.

Fig. 3. Zero Gravity. © Kitsuo Dubois, France

With Dubois, presence is hidden in time, it's the time itself, the time of the body awakening to a new structure of forces or absence of forces. Flesh and muscles and bones are suddenly submitted to an unexpected and never experimented before behaviour.

Let's think of the differences between these works and those of Marey or Muybridge. The chronophotographical works of Marey or Muybridge give us no sense of "presence" because the representations of the bodies are immerged in a time entirely stuctured by space.

With Carlier's work, the sense of "presence" comes from the fact that the representations of the body are placed in a space entirely structured by time.

With Dubois' work, the body is also placed in a mobile space: the very structure of the properties of space are modified in time (during the few seconds of zero gravity).

Conclusion

We need a real effort of refinement in our understanding of the different "levels" of presence, and of the different "levels" of representations they are associated to.

The blurring between presence and representation will continue to increase, leading us towards very confusing mistakes.

It's not only art that might be affected by this confusion, but also our understanding of the reality.

We are led to ask ourselves what could be a good criterium of "presence", or a good criterium of "reality".

As with the examples shown before, we suggest that such a criterium could be found more in time than in space.

Space appears to be a kind of a "matter" ready to be "shaped" by time. Time is the ultimate condition to exert one's thought, to exert one's own meditation over the essence of presence.

But time is only a condition of exercise. It is not a criterium of judgment.

In the context of virtual communities, we'd like to suggest that the criterium of "reality" should not be found in representations, however realistic. Since a virtual community is a socialized space, its "reality" is linked to what makes this social construction survivable, hence the criterium of "common good". The "reality" of a virtual community is not its image but its finality, itself measurable by its capacity to fulfill goals compatible with the common good.

This is where art joins ethic.

References

1 http://www.ina.fr/INA/Recherche/TV/TV.fr.html

2 Husserl E (1986) Die Phänomenologie und die Fundamente der Wissenschaften. Text see "Husserliana", Vol. 5, Meiner, Hamburg

3 Dubois K (1994) Dance and weightlessness: Dancers' training and adaptation problems in micro gravity. In: Malina R (ed) Leonardo, Vol. 27 (1), pp.57–64

Telematic Art

Maria Grazia Mattei

A little known chapter of contemporary art regards the movement begun during the seventies that was created around the use of the new communication technologies. The computer, the telephone, but also satellites, slow video and the telefax over a short time became instruments to develop performances, and exchanges of information and data (images, sound, texts) from one part of the world to another in real time in a creative way. In an analogical period information travelled more slowly and action was much less immediate. Nevertheless, thinking through those experiences again carefully, they were so rich in intuition that they almost seem more current today in the digital era, in the network era of numerical satellites, of the information superhighways than at that time.

The history of Telematic Art, or Communication Aesthetics as it has also been defined, constitutes a significant wealth in order to understand not only the current evolution of information and telematic technology and the reason for their success, but also in order to reflect on a series of still valid concepts, such as the relationship between time space, interactivity in the creative process of communications, collaboration, democracy, transparency in the network of human relations, etc. The future vision of a better "human condition", telematics itself understood as the aesthetic extension of participation in social life. These are just some of the ideas pointed out and developed by Roy Ascott beginning in the seventies while others for the entire first decade insisted many times on the necessity to develop the real possibilities for a total, immediate and extensive communication system (see Robert Adrian X).

With the advent of the new technologies and digital media a new phase in aesthetic research has also begun. The use by artists of new instruments and forms of telematic communications have as a first real consequence the mutation of the ways for the creation and enjoyment of works of art.

The transformation of the forms of presence, the delineation of a new phenomenology of time-space, and the creation of a new kind of event like the formation of a new type of hybrid energy between man and machine are among some of the main points that critiques like Gillo Dorfles, Fred Forest, Mario Costa, Filiberto Menna and artists like David Rokeby, Giovanna Colacevich, Robert Adrian X, etc. debate and provoke often in their meetings.

Up to today this has been an artistic exercise which, like other avant-garde research, has put the accent on the creative process between artist and audience, rather than on the work of art itself. "The aesthetic specificity of the Art of Telecommunications and the Art of Communications," declares Frank Popper in his book 'L'art a l'age electronique', "in its inseparable togetherness from the technological specificity regards both its creation and its reception, which are closely connected here more than in other art forms..."

The Telematic Art has radicalized these aspects within a debate on art and has opened, with great anticipation on the present, themes of psychophysical mutation, for example, the right to a wide spread creativity, the development of a pluralistic culture, virtuality of images and objects, and cultural mutations. Today these themes are just as urgent, because, with the technological conditions changed, we are all involved, and not only in a narrow avantgarde, to live immersed in a world made of inedited relations between protocols and rules still to be established.

The first experiences of Telematic Art go back to the first half of the seventies when communications satellites were diffused. Artists used this channel to relate different media and to carry their messages long distance.

In 1976 Douglas Davis created Seven Thoughts, a performance that gave the possibility to launch free thoughts throughout the whole world in real time through Intelsat from the Houston Observatory. The first experiences with a satellite tried to reach a dimension and global distribution of thought, in a certain sense, by realizing almost the first communications network alternative to the official media. Instead others also forecast interactive systems like Send/Receive (1977). Organized by the Center for New Arts Activities of New York and by Art Com La Mamelle of San Francisco, it planned 15 hours of interactive transmissions between the two cities. In the same period Satellite Ars Project was the first linked interactive performance between the two coasts. Realized by Kit Galloway and Sherrie Rabinowitz with the Nasa Goddard Space Flight Center in Maryland and the Educational Television Center in Menlo Park, California, the performance consisted of a dance event shared and performed by two different groups of dancers far from each other. In Europe in 1977 at Documenta VI in Kassel, Germany, Nam June Paik, Joseph Beuys and Douglas Davis created their own television performance live via satellite sent to more than thirty countries.

From 1975 Sherrie Rabinowitz and Kit Galloway focussed their collaboration on the development of new models of enjoyment where the performing arts were combined with public participation in sophisticated interactive telecommunication structures. They were the ones that gave birth to the current Electronic Cafe International in Santa Monica that is often spoken about. The project derived from an experience in 1984 commissioned by the MOMA in Los Angeles for the Olympic Games. At that time five places in L.A. diverse in ethnic and cultural identity were constantly linked giving birth to a strange Arts Festival. The activity of the ECI today occupies a relevant position in the arts and interactive telecommunications field both for the activities carried out in the past and for current events planned today. The ECI boasts of collaborations with the most important museums and arts centers like La Villette in Paris, Documenta in Kassel, etc. The main events of the ECI usually consist of multimedia links with artists, institutions, museums, galleries, festivals or other International Electronic Cafes by creating Teleperformances and combining various disciplines such as poetry, music, dance and the visual arts thanks to the new technologies. The ECI has become a model "copied" in various cities throughout the world. This is the present story but already with Hole in Space (1980) Kit Galloway and Sherrie Rabinowitz made quite a fuss. The two California artists planned and realized this event by linking New York and L.A. via satellite. Cameras and monitors were placed on the inside of the show windows of two stores chosen in the respective cities in such a way that the public that casually passed could communicate live with images and sound without being prewarned. The event lasted for three evenings and people in great numbers, evening after evening, crowded around the windows on the street.

In those years in Europe Roy Ascott began his activity with a computer conference dedicated to an artistic project by using the Planet Network, one the first development prototypes of the large Internet network.

In 1982 World in 24 Hours was organized instead by Robert AdrianX for the Festival Ars Electronica in Linz. The project linked 16 cities in three continents for 24 consecutive hours. The artists were invited to WienCouver also organized by Robert Adrian in 1983 in collaboration with Hanc Bull (Vancouver), Helmut Marc (Wien), Reinald Schumacher (Berlin) and Marcin Krzyzanowski (Warsaw). It was a low cost performance event but very significant for its performance climate of the time with music and teletransmitted images by telephone cable. Robert Adrian can be considered one of the pioneers in telematic art. His work tends to analyze profoundly the nature and potentiality of telecommunications understood as an artistic instrument. The problems that are brought out in the use of telecommunications can be broken down into two aspects: the access given to artists to the mass media and the hi-tech demonstrations of technologies. "Communications have to be refined, not the mean," according to Adrian, "also the use of the technology should be made accessible to not only the populations of the western, industrialized world, but also on a universal scale in a global dimension, otherwise it is just literature".

The aesthetics of AdrianX can be summed up in the project ZERO, the art of being everywhere, a cultural initiative created in Styria, Austria which has as its aim the creation of infrastructures that can help and promote the use and spread of the new electronic and digital media. From Adrian's school various other groups in Austria have sprung up. The most active one is certainly the Atelier Graz. Its recent Winke Winke installation is among the most significant works presented in the most important international exhibitions (The Venice Biennale '93, Siggraph '94 in L.A., The Interactive Media Festival '94 in L.A. and The Ars Electronica Linz '94). It is a remote controlled robot that transliterates written messages from visitors in nautical code, an obvious symbolism of communications mediated by technology.

In 1983 La Plissure du Text by Roy Ascott, is a telematic event realized at the Beaubourg in Paris for Electra, one the the festivals which in that period was dedicated frequently to the scene of electronic art. It was based on the collaborative realization of "stories" in order to demonstrate the potentiality of the collective author in a dimension of "global" communications. The participants were Robert

AdrianX (Wien and Vancouver), Bruce Breland (Pittsburgh), Eric Gidney (Sidney), Norman White (Toronto), Carl Loeffler and others scattered around the world.

Later in Linz through Aspects of Gaia (1989), an interactive installation with images, texts and sound, Ascott carried forward the reflections on the collective author, which was very successful. By starting from the analysis of network telecommunications systems and from the information contained in the data banks and utilized as raw material, Ascott believes, as his principle thesis by drawing inspiration from Wiener and Ashby, the possibility that through a cybernetic approach to art and a collective creative work project, it is possible to reach superior levels of knowledge. The creation of a "universal creative network or a universal network creativity" is the concept that Ascott has followed in various projects. Planetary conviviality, collective participation, the use and choice of texts as a less costly and more manageable language, and the idea of the informed telematic network as a metaphor of human interconnections are some of his themes.

In the year of Orwell (1984) the most famous videoart artist, Nam June Paik, launched Good Morning, Mr. Orwell, a television transmission via satellite created, realized and directed by Paik himself via satellite with Paris, New York, Seoul, Cologne and San Francisco.

This was a transcontinental artistic program with links, interventions and interviews in real time and prerecorded, completely remote controlled and manipulated by Paik himself, who like Big Brother, intervened electronically on the emission from remote direction consoles. The objective of Nam June Paik was to mix continents, languages, artistic cultures and different societies live. On the transmission international avantgarde artists and musicians, such as Laurie Anderson, Peter Gabriel, Allen Ginsberg, Urban Sax, John Cage, Merce Cunningham, Joseph Beuys, Philip Glass and others participated.

"Good Morning, Mr. Orwell" will be remembered as the first artistic program transmitted via satellite by public television which conceded the privilege of a live broadcast to an innovative, experimental and explosive creativity, such as that of Paik in an attempt to reach a vast audience and dialogue on a global scale.

In Italy experimentation in the area of telecommunications began in 1984 in Pavia with the show Arte e Nuove Tecnologie (Art and New Technologies) which offered the public artistic videoinstallations as well as live links via fax. The Venice Biennale in 1986 instead dedicated an entire section to the theme by creating a special place. Network Planetario, edited by Ascott, Adrian, Trini and Forest was the title of the project which revealed, with great energy, the new innovative potentialities of new research based on the new technologies to the general public. The analysis task and spreading around these themes was done nevertheless with great competence by the two groups. The first one was the Aesthetics of Communications, an association founded and directed by Mario Costa at the University of Salerno together with Fred Forest in Paris. On several occasions they have organized debates, events and important publications. Tempo Reale is the second group operating in Italy and it began in 1986 in the village of Calcata initiated by Giovanna Colacevich and Giuseppe Salerno with the opening of a gallery and various artistic laboratories dedicated to the research for new forms of communications. The most significant events were held in various galleries that made connections with Kenya, Japan and the United States as, for example, "Telefax, Rome ... Nairobi" (1988) and "Foto di Gruppo" (1989).

The new electronic digital media toward the end of the eighties constituted the investigation field on which the attention of the Ponton Media Lab in Hamburg, which realizes events aimed at overcoming the traditional division of tasks and competencies between artists and technicians, was focussed.

The Ponton European Media Art Lab/Van Gogh TV is an Austro-German project created in Linz in 1986. In 1989 a laboratory was installed in Hamburg with the aim of favouring the collaboration between artists, engineers, theoreticians and technicians for the realization of new interactive technological systems by privileging the communicative, social and artistic side of the new media.

Ponton and the various work groups connected such as Van Gogh TV, Minus Delta and Van Gogh Radio have realized a lot of works for important international shows and expositions like Documenta (1987), Ars Electronica in Linz (from 1980 to 1990) and again in Kassel in 1992 where they realized Piazza Virtuale, one of the first interactive television experiences. Through various connection means (telephone and satellite) interactions among various users in the European area were created on the home screen. The metaphor utilized was the Town Square where people meet to discuss and interact urged by games, interviews, designs and music especially prepared by Van Gogh TeleVision. Experimentation at the Ponton Media Art Lab has made great advances, so much so that from this laboratory the most interesting information has recently come out regarding the development and diffusion of three-dimensional graphic interfaces for multimedial network services. The virtual environment and the communication means in a space that belongs to the memory of a computer but that can be shared by one or more persons is the central theme of these days.

Home of the Brain realized in 1992 at the Art+Com in Berlin by Monika Fleischmann was the first great metaphor of televirtual communications, immersible for a city that is more and more mediatic. However, it was above all the Cluny experiment organized by Imagina Festival in 1993 that struck the collective imagination when two persons physically some distance apart found themselves together inside the improbable and symbolic space of the antique Abbey destroyed and rebuilt in virtue of the representational power of the computer. Today with the phenomenon of network access the most divulging and experimental phase seems to be over, linked to the themes of the collective author or the right of access. Artistic and experimental research have moved rather to the themes of interactivity in virtual environments, long distance collaborations, themes of participative democracy, political aspects linked to the development of the information society with installations, that challenge the codes inside which information technology seems to be confined in order to give weak signals bound to have a great resonance.

Image/Speech Processing Adopting an Artistic Approach – Toward Integration of Art and Technology

Ryohei Nakatsu

1. Introduction

In this paper, the possibilities that might emerge by combining image/ speech processing technologies and art are discussed. Generally, for engineering technology tends to open at the forefront and eventually goes far away from human factors in the name of "high technology". In contrast, art expresses the deepest parts of humans such as emotions or senses. In other words, art and technology seem to be like oil and water.

In ancient times, however, each person could make earthenware or stone tools. Some craftsmen could also draw pictures in places like the "Altamira Cave". Therefore, in ancient times, a lot of people were both an engineer and an artist. In modern times, on the other hand, the gap between art and technology has developed with the rapid progress in science and its specialization.

Even at the present time, we can find many cases in which art is being combined with technology in daily life. In the case of clothes, for example, artistic elements such as design have been becoming important, although clothes were originally industrial products to protect the human body. In the case of cars, not only their functions but also their design is important. Even electric products for domestic use, such as vacuum cleaners or washing machines, require good designs.

On the contrary, for high-technology products represented by personal computers, design aspects such as usability have not been considered yet. Moreover, there are many cases in which design has been slighted in high-technology machines, such as with speech/character recognition equipment. In these situations, people become fearful of the personal computers or high-tech products and, therefore, tend not to accept these high-tech machines. In the future with an aging society, it is anticipated that these tendencies will become more and more remarkable.

In the field of communications, the development of new communication systems and services for the next century is expected by utilizing the high technology called "Multimedia". However, we cannot deny that there is some anxiety that our future society, to be full of high-technology equipment, will lack human compassion and therefore, will be gloomy. The reason for this is that recent technology has been advancing in a direction that ignores the human senses and emotions.

Yet, we think it is important to develop services and systems while considering the human senses and emotions. For this reason, we believe it is necessary for engineers to work together with people who can handle these human factors, such as artists. Based on this point of view, in our research laboratories, we are carrying out research, aimed at new communications technologies, based on collaboration between artists and engineers. In this paper, basic concepts and examples of such trials will be stated.

2. Communications and Image/Speech Processing

2.1 Intellectual Activities of Human Beings and Image/Speech Processing

Handling the intellectual activities of human beings is Artificial Intelligence (AI)'s main subject. Among various kinds of intellectual activities we focus on the human communication functions. The main reason is that general aspects of intellectual activities are expressed in communications. Another reason is that results from research on communications technologies can be applied widely to our lives and societies. In this area, so far, engineers have been concentrating their research on robots or computer agents that have functions to communicate with human beings. The major part of this research has been emphasizing only the verbal aspects of communications. For example, speech recognition has been aiming at extracting basic meanings, that is, verbal information. This is the same situation as for character recognition, especially handwritten character recognition; research on algorithms has been done to extract information on meanings from handwritten characters. However, it has been recognized that in daily life, the transfer of emotions and senses, that is, non-verbal communications, also plays an important role. For example, speech includes speaker-related information and emotions-related information in addition to verbal information. In speech recognition, however, non-verbal information has been ignored and treated as noise.

Creating human-like computer agents or characters requires the research and development of technologies concerned with non-verbal communications. Agents adopting such technologies may be able to have hearty communications with human beings.

2.2 Communication Model

Figure 1 shows a human communication model. It should be noted that this model has a similar construction to the human brain. In the outer layer, which corresponds to the new cortex in the brain, there is a layer that controls communications based on the use of a language. Researchers in the AI field have been studying the mechanism of this layer.

As stated above, speech recognition and character recognition are typical examples in AI research. Other key research areas in AI are planning and inference. In the field of speech recognition, research has been done for many years on algorithms that can achieve a high recognition performance by handling only the

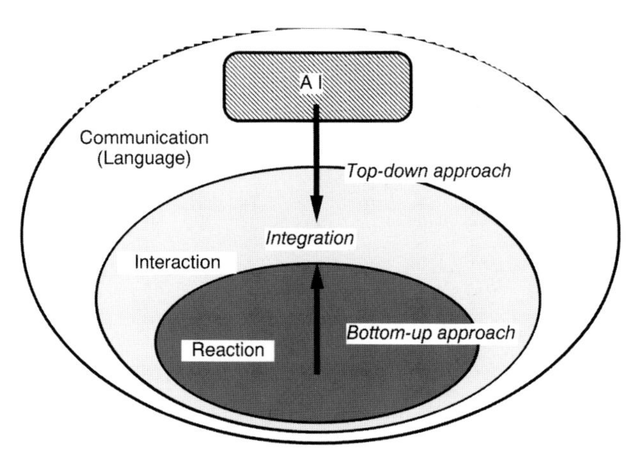

Fig. 1. Subsumptive communication model

logical information included in speech. However, it can be said that this is merely a partial point of view which neglects essential problems. As stated before, logical information is only a part of the whole information that constitutes speech. Other rich information, like information on emotions or senses are also included. Such information are considered to be created at the deeper level layers, that is, the Interaction and Reaction layers indicated in Fig. 1.

The interaction layer controls actions to maintain the communication channels, like nodding, controlling speech rhythms, or managing changes of speech turn. This layer plays an important role in achieving smooth human communications. Under this layer is the Reaction layer, which controls more basic actions. Examples of actions include turning one's face toward the direction from which a sound had come or closing one's eyes upon suddenly sensing a strong light. Such functions were obtained in ancient times when human beings were still uncivilized. In other words, functions common to all animals are controlled in this Reaction layer.

Thus, not only the handling of logical actions and information but also the handling of functions at deeper layers plays an important role in human communications. These functions create and understand non-verbal information like emotions and senses. The reason why efficiency in speech recognition has so far been limited is because such essential information has been neglected as noise. Therefore, in order to understand general human communications functions including the sending-receiving of emotional information other than logical information, it is necessary to research the mechanism of the Interaction layer and Reaction layer and to integrate the results with the functions of the Communication layer. By doing so, agents with human-like behaviors can be made.

3. Approach Aiming at the Integration of Art and Technology

In the previous section, the necessity of studying the action mechanisms of the deeper level layers in human communications was explained. This section proposes the idea of integrating this technology and art.

As stated before, in the engineering field, research is being done targeting the handling of logical information in human communications. As the research advances, however, it is becoming clear that the mechanisms of deeper level communications, like communications based on emotions or senses play an essential role in our daily communications. We can call such communications "Interactions" or "Reactions". It is, therefore, inevitable to be able to handle information on emotions and senses, which had not been handled in the engineering field up to now. On the other hand, artists have long handled human emotions and senses. Therefore, further development is expected by having engineers collaborate with artists.

Art too has seen a notable movement recently. This is due to the emergence of a field called Interactive Art. The important function of art is to have an artist transfer his/her concepts or messages to an audience by touching their emotions or senses. In the long history of art, this means of communications has been refined and made sophisticated. However, it cannot be denied that in traditional art, the flow of information in communications has been one-way, that is, information is transferred from the artist to a passive audience.

With Interactive Computer Art, the audience can change expressions in art works by interacting with them. That is, the audience provides feedback to the various art works and this consequently enables information to flow from the audience to the artist. Therefore, in Interactive Art, information flow is both ways, that is, true communications is achieved. A comparison of information flows between traditional art and Interactive Art is illustrated in Fig. 2.

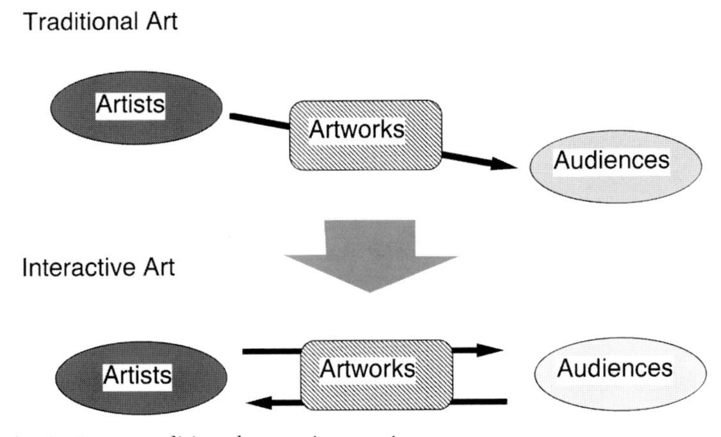

Fig. 2. From traditional art to interactive art

At the same time it should be pointed out that this Interactive Art is still developing and that sometimes interactions remain simple, like causing a change by pushing a button. Therefore, it can, in some cases, be useful for Interactive Art to adopt image/speech processing technologies to raise primitive interactions to the communications level.

For this aim, from an engineering viewpoint, collaboration with art is required to give computers human-like communications functions. From the art side, adopting new technologies can help to improve the current Interactive Art from the level of interaction to that of communications. As both approaches share the same target, the time is ripe for collaboration between art and technology to progress.

Moreover, these two approaches not only contrast but also supplement each other. In the case of engineering, observing phenomena by an analytic approach and constructing a model based on the observations are mainstream. In the case of art, on the other hand, approaches are based on the intuition and sense of artists. When we make for example an agent with a human face, engineers try to create a face as close to a human's as possible by utilizing CG technology. In this case, the more the agent looks like it has a real human face physically, the more we notice slight differences and sometimes have very unnatural feelings towards the face. Artists, however, can express substantial characteristics of a human face by drastic deformation. Though physically such a face is different from a real human face, we can obtain a familiar feeling to such a face. By integrating these contrastive and supplementary creation methods of faces, a more natural and familiar agent can be expected.

4. Examples of Approaches Integrating Art and Technology

In our laboratory, based on the above idea, we started to employ artists in the Interactive Art field from last year and began new attempts to carry out research based on collaboration and joint activities between artists and engineers. Several of our attempts are classified as follows (Fig. 3): (1) Only artists: Artists themselves have engineering knowledge and the ability to produce art work by adopting new technologies. (2) Artists and engineers: Artists present a new concept and engineers provide technologies to realize it. Thus, there is collaboration. (3) Engineers and designers: Engineers present a whole concept and Artists produce the art part of it. Thus, there is collaboration. (4) Only engineers type: Typical engineering research.

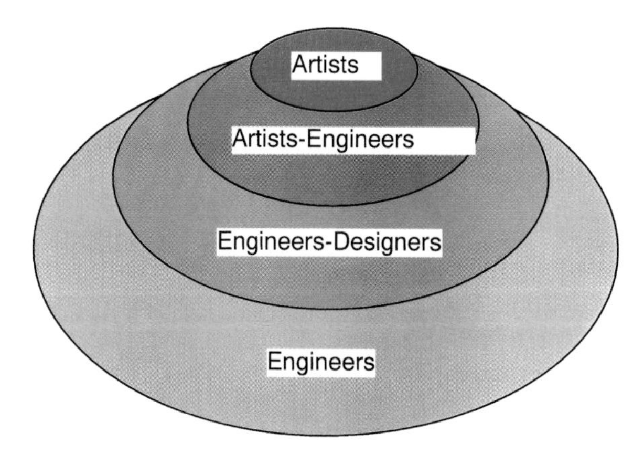

Fig. 3. Collaboration between engineers and artists

In the following, some examples of research activities in our laboratory are described.

4.1. Interactive Communications Environment "MIC Exploration Space"[1]
The environment plays an important role in human communications. In most cases, the environment is fixed and we have to choose a place where communications are achieved depending on the communication content, like a meeting room for business, a coffee shop or a restaurant for talk with friends, or a living room for a relaxing time with family members. Now, however, it is possible to create various kinds of environments adequate for communications by applying image and speech processing technologies. CG technology is the key among various kinds of technologies to create an environment that changes in real time depending on the content of communications. We call such a communications environment an "Interactive Communications Environment."

The "MIC Exploration Space" has been produced by Christa Sommerer and Laurent Mignonneau as a concrete example of an interactive communications environment. They are artists in the field of interactive art and have been producing virtual plants and virtual creatures that interact with human beings since the beginning of the 1990s. Two representative examples are explained below.

"A-Volve"[2]: The theme of this art work is interaction between virtual creatures and human beings. Virtual lives created according to inputs on a touch screen are projected onto a screen installed under a pool and the audience can interact with the virtual lives by hand movements done in the pool.

"Trans Plant"[3]: The theme is interaction between virtual plants and human beings. On a screen installed in front of an audience, virtual plants and images of the people are projected. The virtual plants grow according to the movements of the people. Therefore, the audience can create an environment of their own.

In the MIC Exploration Space, Trans Plant's basic concept of interaction between humans and virtual plants is extended to communications between humans or between humans and the environment. In Fig. 4, the whole image is shown. Suppose that two same systems are placed at distant places and they are connected through telecommunication lines. In front of the two screens, two persons are standing. Their images are taken by cameras that are set just above

Fig. 4. MIC Exploration Space (Christa Sommerer & Laurent Mignonneau for ATR MIC Lab, 1995)

Fig. 5. Interaction in "MIC Exploration Space" (Christa Sommerer & Laurent Mignonneau for ATR MIC Lab, 1995)

the screens and only human images are extracted from the background. At the same time, cameras installed above the two persons recognize their standing positions and their simple movements, and CG images of virtual trees, grass and flowers are created according to the recognition results. Overlapped images of the two persons and CG plants are projected on the screens. Consequently, each person can feel as if he/she is meeting the other person, who is at a remote place, in the virtual space. Also they can feel, when their environment changes based on their communications, as if they can create their environment based on their communications. In Fig. 5, a situation where this system is used is shown.

4.2. Emotional Agent "MIC & MUSE"[4,5]

In human communications using voices, emotion plays a very important role. Sometimes, information on emotions is more essential than the logical information included in speech. This can be confirmed from the fact that babies start to recognize emotional information before they recognize information in their mothers' voice. In the case of adults too, we can recognize what other people want to say at a deeper level by integrating information on meanings and emotions included in speech. This is the key for communications to proceed smoothly. Unfortunately we have to say that, in the field of AI so far, focus has been on recognition of only meaning information and emotions have been neglected as noise. In order to create an agent with human-like behaviors, therefore, it is necessary to add functions enabling it to recognize emotions and to react to them.

"Neuro Baby"[6,7] was produced by Naoko Tosa according to this idea. Neuro Baby is a computer character that is capable of recognizing four emotions included in speech and reacting to them by changing his facial expressions. Based on her experiences in developing and exhibiting Neuro Baby, she also produced "MIC & MUSE" in cooperation with engineers at our laboratory last year. In comparison with Neuro Baby, MIC & MUSE have the following improvements.

(1) A better ability in non-verbal communications: MIC is a character that reacts to emotions involved in speech just like Neuro Baby. He can recognize eight emotions (joy, anger, surprise, sadness, disgust, teasing, fear, and normal) to Neuro Baby's four. Emotions communication, however, is only one aspect of non-verbal communications. In our present study, therefore, we included another kind of non-verbal communication: communication based on music. In addition to MIC, MUSE, which communicates with humans by music instead of voice, was produced. MUSE relates simple melodies input by a xylophone with emotions (e.g., ascending scale to joy, descending scale to sadness), and responds to them with piano melodies.

(2) Improvement in reaction patterns of a character: A CG image of a full-length portrait was created and the character's emotional reactions were expressed by whole body reactions as well as facial expressions.

(3) Improvement of functions for emotion recognition: By adopting the following methods and technologies, functions for emotion recognition were improved.

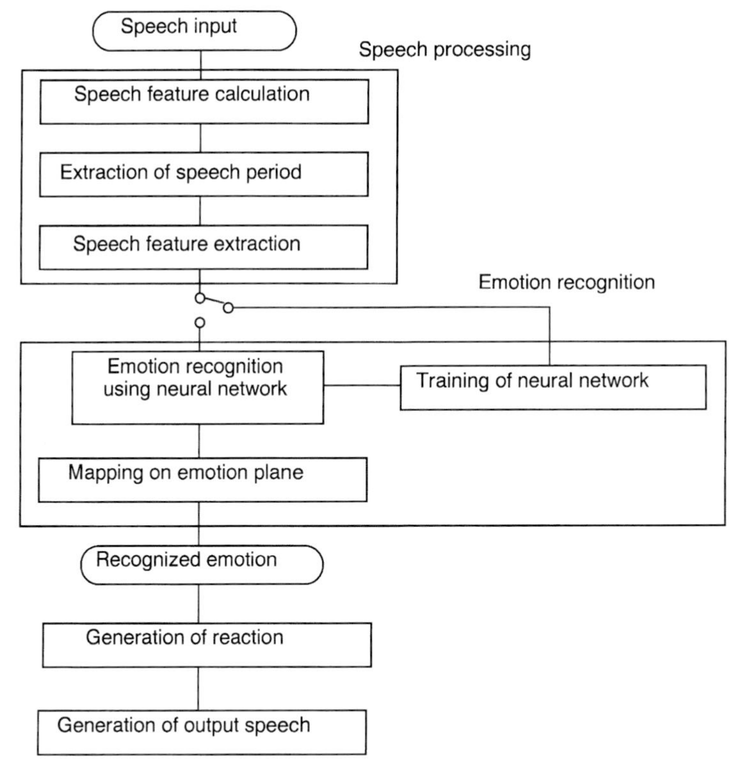

Fig. 6. Block diagram of the processing flow

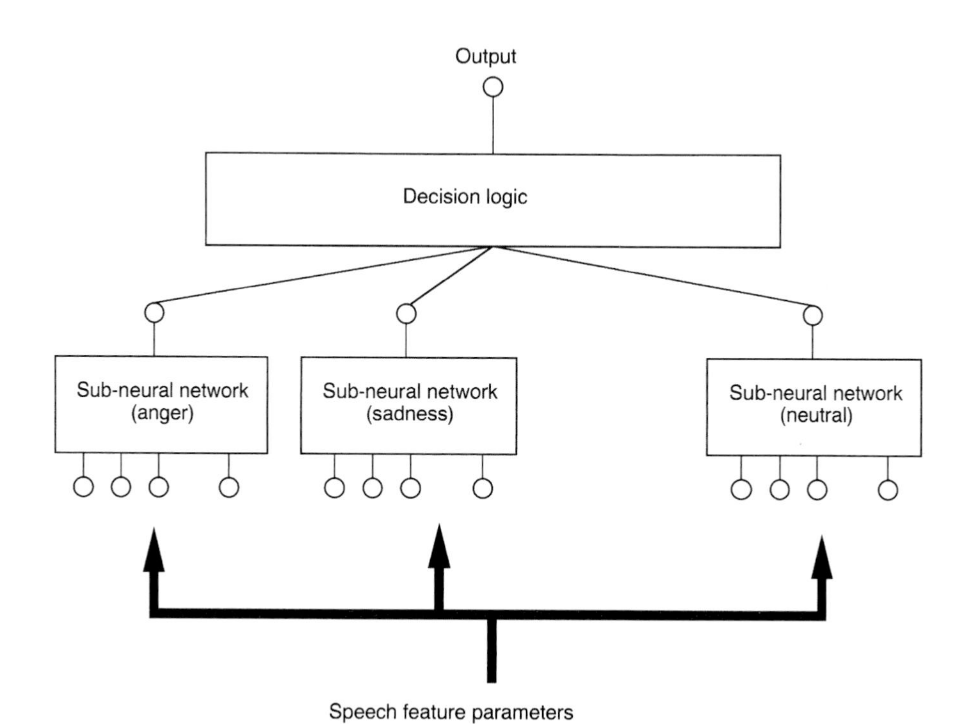

Fig. 7. Configuration of emotion recognition part

Fig. 8. An example of MIC's emotional expression

To improve the emotion recognition capability of MIC & MUSE, a combined-type neural network architecture was introduced for emotion recognition. Eight neural networks that corresponded to each of the eight emotions were prepared and emotion recognition was achieved by feeding feature parameters into these eight networks in parallel.

By using speech data, that is, various kinds of phonetically balanced words uttered by many speakers, as training data, speaker-independent and content-independent emotion recognition became possible.

The whole processing flow is shown in Fig. 6 and the construction of the emotion recognition part is presented in Fig. 7. An example of MIC's representative reaction patterns is given in Fig. 8.

4.2. Virtual Kabuki[8]

Facial expressions play an important role in human natural communications. We communicate with other persons smoothly by recognizing their emotions through their facial expressions and also by expressing our emotions through our facial expressions. In order to realize an agent with a human shape in a virtual space, therefore, a technique is required to extract facial expressions from its image in real time and to reproduce them with a three-dimensional face model. For this objective, we studied the real-time recognition of facial expressions and reproduction technologies. As an example of applying this technique to the creation of a human-like agent, we examined the reproduction of a three-dimensional face model of a Kabuki actor.

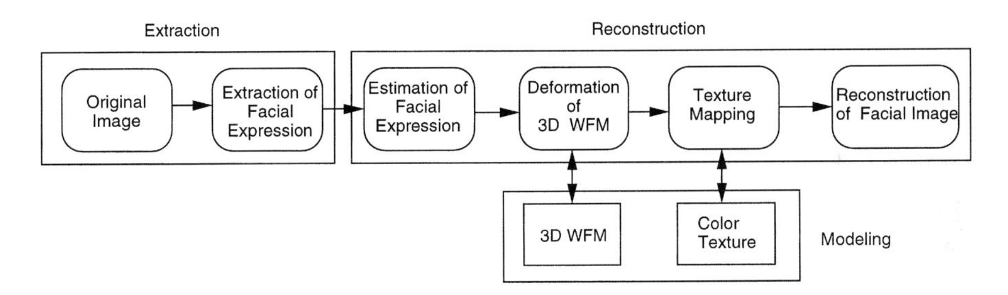

Fig. 9. Flowchart for reconstruction of facial image

The flowchart for recognition of facial expressions and creation processing is shown in Fig. 9. The recognition of facial expressions and creation processing system consist of three parts: extraction of expressions, face reconstruction, and face modeling. The face model must be created beforehand; a three-dimensional model of the face is created in the form of a wire frame model. With this wire frame model, the facial shape is made similar to an assembly of small triangular patches and the color texture of the face is rendered on these triangle patches. In order to extract facial expressions in real time, the subject puts on a helmet and a small video camera attached to the helmet takes images of the subject's face. If the position or direction of the head changes, the helmet follows these changes and always extracts facial images stably. Next, DCT conversion (discrete cosine conversion) is carried out on the obtained images by the camera and changes of facial compositions, such as the eyes or mouth opening/closing, are extracted. Information on the changes are reflected to the transformation of the three-dimensional model and the extracted facial expressions are reconstructed as facial expressions of a Kabuki actor. An example of a reconstructed Kabuki actor is shown in Fig. 10.

Fig. 10. An example of Kabuki actor's face model

Of course, a key point of this system is, technically, the extraction of facial expressions in real time. At the same time, the transformation technique, which extracts the expressions and reproduces them as a Kabuki actor's expressions, is essential. Note that an artist creates the Kabuki actor's face model to add an artistic touch. By adding such an artistic element, anyone can transform himself into a Kabuki actor.

By extending this idea, the possibility of new entertainment, such as the combination of a role playing game and a movie, where someone enters into a virtual space and experiences various stories, can be born.

5. Conclusion

In early times, art and technology were the same thing. However, as technology started to target the support of human life materially and art began supporting us from a spiritual aspect, they started to go in different directions. Now, at the end of the 20th century, as a new direction for technology, the necessity to support humans spiritually is being recognized. In the world of art too, a new movement is emerging. By utilizing new technologies, many artists are trying to create new art and provide new methods to express the human spirit. Therefore, the possibility that technology and art can collaborate or unite again is emerging.

In this paper, the possibility of new technologies which will be developed by integrating art and technology was discussed by referring problems current technology is facing. It was also stated that, by combining AI research, such as processing technologies for image and speech, with artistic approaches, the possibility exists that fundamental technologies as well as systems and services for new communications can be created. As examples of this new direction, some projects have been introduced that are being done at ATR Media Integration and Communications Research Laboratories. These projects have just started but we highly await their results. Detailed progress reports will be given on other occasions.

References

1 Sommerer C, Mignonneau L (1996) MIC Exploration Space. SIGGRAPH Visual Proceedings. The Bridge, pp. 17
2 Sommerer C, Mignonneau L (1997) A-Volve – an evolutionary artificial life environment. In: Langton C, Shimohara K (eds) Artificial Life V, MIT, pp. 167–175
3 Sommerer C, Mignonneau L (1995.2) Trans Plant, Imagination, Tokyo Metropolitan Museum of Photography, Chapter 2
4 Tosa N, et al. (1994) Neuro-Character, AAAI '94 Workshop, AI and A-Life and Entertainment
5 Tosa N, et al. (1995) Network neuro-baby with robotics hand, symbiosis of human and artifact, Elsevir Science B.V., pp. 77–82
6 Tosa N, Nakatsu R (1996) The esthetics of artificial life, A-Life V Workshop, pp. 122–129
7 Tosa N (1996) The esthetics of recreating ourselves, SIGGRAPH'96 Course Note on Life-like, Believable Communication Agent
8 Ebihara K, et al. (1996) Real-time facial expression detection and reproduction system – Virtual KABUKI System –, Digital Bayou, SIGGRAPH'96

2. Scientific Visualization

What Can an Artist Do for Science: "Cosmic Voyage" IMAX Film

Donna J. Cox

Abstract

A brief history of visualization and science is related to contemporary supercomputer visualizations. During the Renaissance period in western history, collaboration among artists and scientists was encouraged as a working model. Today, collaboration plays a major role in various forms of art such as filmmaking, computer animation, and scientific visualization. A "Renaissance Team" is a group of specialists who collaborate to contribute their expertises in solving visualization problems. The making of the "Cosmic Voyage" IMAX film is an example of global multidisciplinary teamwork to create the first narrative large-format movie with greater than 14 minutes of complex, high-resolution scientific visualizations for special effects. The "Cosmic Voyage" was nominated for Academy Award in documentary short subject category.

Professor Cox is the Associate Producer for Scientific Visualization and the Creative Director for the NCSA/PIXAR segment of the "Cosmic Voyage." She orchestrated a global "Renaissance Team" of artists and scientists from two national centers for supercomputing, PIXAR Animation Studios, Princeton University, University of California at Santa Cruz and Santa Barbara Studios. A new technology was invented to choreograph the computer graphics camera using voice-controlled virtual reality in the CAVE(TM).

In addition, a collaboration in a university environment is described. Cox's former students received a nomination for an Academy Award for short animated film, in 1996. Arguments are presented why the "Renaissance Team" model is an excellent collaborative method for art/science research as well as education.

Collaboration in Art und Science

During the Renaissance in western culture, many scientists and artists collaborated to visually study natural phenomenon. Many Renaissance artists came to believe that the visual observation of nature could reveal its hidden laws. This philosophy set the stage for the scientific revolution that was to follow. Galileo developed a scientific method to visually and objectively gather data; to form theories; and to encourage peer review. Within 100 years, Newton had developed calculus as a mathematical tool to encode and predict natural phenomenon. Mathematics was adopted as a method for scientific research and a method to predict the future.

In the 20th century, computational science began a new era in scientific research where large supercomputers are used to compute mathematical models of theories. Artists help create visualizations of these models by using computer graphics. Teamwork and collaboration is extremely important to most of this large-scale science, especially for research that involves visualization techniques.

Filmmaking is a collaborative artform. The primary reason for this organized approach is because of the complexity and technological requirements involved in filmmaking. For the same reasons, films and most computer animations are products of teams of workers. Likewise, the best scientific visualizations are a result of teamwork.

The Making of "Cosmic Voyage"

The IMAX movie "Cosmic Voyage" was released August 1996 at the Smithsonian National Air and Space Museum (NASM) in Washington DC, and now is playing around the world. This movie has many technical firsts including the fact that it contains more computer graphics imagery (CGI) and scientific visualization minutes than any other narrative IMAX movie. Each IMAX motion picture frame is about 10 times the area of a standard 35 millimeter film frame. The typical IMAX theater screen is 70-foot wide. Thus, this movie represents a major CGI challenge because the slightest error in the 4000 pixel resolution is magnified by the scale of the screen pixels.

Fig. 1. Colliding Galaxies Visualization. © Donna Cox and Bob Patterson, UIUC

The Associate Producer for Scientific Visualization and the Creative Director for the PIXAR/NCSA segment of the film orchestrated a major collaboration among PIXAR, Princeton, University of California at Santa Cruz, National Center for Supercomputing Applications (NCSA), Santa Barbara Studios, Electronic Visualization Lab – UIC, and San Diego Supercomputer Center (SDSC) to produce scientific visualizations using data driven supercomputer visualizations.

"Cosmic Voyage," is funded by NASM, Motorola Foundation, and National Science Foundation, and was produced by Cosmic Voyage Inc. The movie is a visual story about the vastness of the universe: the relative size of things from galactic clusters to quarks. Several minutes of the movie show the birth of the universe to galaxy formation and collisions.

Dr. Frank Summers, Princeton University, created an ultra high-resolution simulation of the condensation and formation of galaxy structures in the early universe. This simulation is 2 million particles per frame, represents 2 billion years, and required a supercomputer running all processors one solid month at NCSA. The result was 120 gigabytes of raw data that shows clusters of galaxies, hot gas, and the interactions resulting in galactic formation.

Another simulation was produced by Dr. Chris Mihos and Dr. Lars Hernquist, UCSC. They modeled a gravitational interplay of 250,000 galactic particles including density and star formation. The simulation shows a double collision and final merger of two spiral galaxies. The San Diego Supercomputing Center donated over 750 hours on a supercomputer for project with the support of the National Science Foundation.

Fig. 2. Colliding Galaxies Visualization. © Donna Cox and Bob Patterson, UIUC

Both of these simulations resulted in huge amounts of data that became extremely difficult to manage remotely. MCI donated time on their ultra-fast network to transfer data across the United States for visualization. The internet was heavily used as the communicaton and data transfer tool for this global collaboration.

Loren Carpenter, senior research scientist at PIXAR Animation Studios, developed a special-purpose particle renderer to realistically render the high-resolution datasets. The software can render multidimensional scientific visualizations; however, this film required "realistic" images that an audience could relate to Hubble photographs. Carpenter's software allows one to control viewing of the supercomputer simulations in a way that is similar to telescopic viewing of observational data: control parameters include exposure, parsecs per units, star magnitude, etc. The Star Renderer also incorporates state-of-the-art rendering techniques such as motion blur which is very important to prevent strobing on large-format film.

Virtual Reality Choreography for the "Cosmic Voyage"

Robert Patterson, NCSA visualization and virtual environment designer, and Cox collaborated with Marcus Thiebaux (EVL) to develop a voice-driven CAVE (TM) virtual reality choreographer: the Virtual Director(TM). This choreographer was premiered at Siggraph '92 Showcase and has been further enhanced to allow intuitive camera capture and editing. The CAVE is an 10-foot cube room with rear-screen projections that allows one to be totally immersed in a stereo projection of the scientific simulations while "performing" the camera motion through

Fig. 3. Using Virtual Director in the CAVE. © Donna Cox, Bob Patterson, Marcus Thiebaux, UIUC

the data sets. Robert Patterson choreographed the simulations for Cosmic Voyage using the Virtual Director.

Erik Wesselak, NCSA software programmer, developed an interface between the massive supercomputer simulation data and the PIXAR Star Renderer. This interface allows separate control of more than 30 parameters and samples and formats data for the Star Renderer. Using this interface, Cox and Patterson aesthetically designed realistic galactic images that move with the dynamics of the supercomputer algorithms.

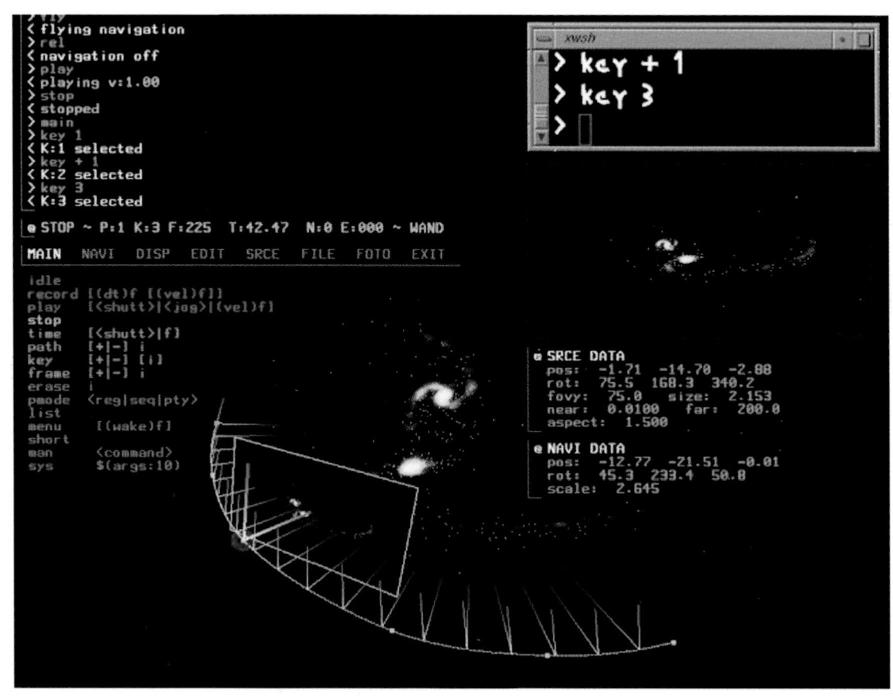

Fig. 4. Menu from Virtual Director. © Donna Cox, Bob Patterson, Marcus Thiebaux, UIUC

The raw data alone was greater than 200 gigabytes, and this data had to be kept on-line to render the final animation tests. The 4096 x 3002 pixel resolution resulted in final IMAX frames that ranged from 10–30 megabytes each. Thousands of these images were stored to disks and transported to Santa Barbara Studios where they were recorded to IMAX film.

Collaboration and Education

Now, with the advent of supercomputing and scientific visualization, there is a demonstrated need for collaborative skills to be coupled to computer graphics education. The Renaissance Experimental Lab (REL) at the National Center for Supercomputer Applications, Champaign, Illinois, USA, was a donated grant from Silicon Graphics. This laboratory is a subset garden of architectures and virtual displays including the CAVE(TM). REL is a unique educational facility are art and science, and it is embedded in an advanced scientific and technology research institute.

Students were organized to collaborate on a production in the REL. "Venus & Milo" is a computer animation narrative centered around a visualization of complex equations and a semi-transparent janitor to a postmodern art museum. The hour-glass female shape of this mathematics resulted in the name "Venus". This animation won the international award from Nicograph for art and entertainment.

After that production, several class members began doing computer animation in filmmaking at ILM, Boss Films, and Warner Bros. Chris Landreth, (now with Alias, Toronto Canada) and Robin Bargar, (now at NCSA), were exceptional students who continued to collaborate after the making of "Venus & Milo." They are now professionals and their latest work, "the end," was nominated for an Academy Award in 1996. This 6-minute computer animated short is an artistic commentary on technology and computer animation.

Conclusion

The extensive use of high technology requires collaboration. Scientific visualization involves a variety of artistic, technical, and scientific expertises. The "Cosmic" Voyage is an excellent example of artists working with scientists to produce graphics that are fantastic and awesome. This type of production would be impossible without global collaboration. Higher education should involve the team approach whenever possible. A large-scale computer graphics animation is an excellent mechanism for this teaching philosophy because of the complexity of the medium.

Glossary

Postmodern – an art and social criticism term that originated in fine art photography and that describes art following modernism, developing during mid-20th century. Postmodern criticism reveals a hypersensitive historical awareness and recognizes the print/electronic media as central to the conduit of culture and suggests that electronic media has created an isolation of the individual from direct experience of reality ; critics observe that art from postmodern era involves a pastiche of recycled styles.

Renaissance Team – a group of specialists with complementary areas of expertise who interact synergetically by pooling their technological, analytical, and artistic abilities to increase the domain of available problem-solving options. For example, teams of artists and scientists in the fifteenth and sixteenth centuries produced classic advances in botany and anatomy; their published works are milestones in the history of science.

References

1 Cox D (1988) Using supercomputer to visualize higher dimensions: an artist's contribution to science, International Journal of Art, Technology and Science 21, pp 233–242

2 Ellson R, Cox D (1988) Visualization of injection molding, Journal of the Society for Computer Simulation, San Diego, California, invited as cover article
3 Cox D (1990) The art of scientific visualization, Academic Computing, Volume 4, Number 6, pp 20–22 and 32–40. Includes cover image.
4 Onstad DW, Maddox JV, Cox DJ, Kornkven EA (1990) Spatial and temporal dynamics of animals and the host-density threshold in epizootiology, Journal of Invertebrate Pathology
5 Cox D (1991) Collaborations in art/science: Renaissance Teams, Journal of Biocommunications 18 (2): (includes back cover image)
6 Cox D (1995) Education and collaboration in an evolving digital culture, Science Visualization in Math Teaching, Chapter 12, Association for the Advancement of Computing in Education (AACE)
7 Cox D (1992) Collaborative computer graphics education. In: Cunningham, S, Hubbold RJ (eds.) Interactive learning through visualization: The impact of computer graphics in education. Springer, Berlin Heidelberg New York Tokyo, pp 189–200
8 Keller PR, Keller MM (eds) (1993) Visual cues: Practical data visualization, IEEE, Los Alamitos, pp 51, 58, 61, 147
9 Cox D (1990) Scientific visualization: Collaborating to predict the future, EDUCOM Review, pp 38–42
10 Cox D (1990) Mapping information, Proceedings, AUSGRAPH 90, pp 101–106
11 Cox D (1991) Scientific visualization: Supercomputing and Renaissance Teams, Proceedings of The 12th New Zealand Computer Conference, pp 157–171
12 Cox D (1991) Beyond visualization: Mapping information, Supercomputing 91 Tutorial Notes
13 Cox D (1990) Beyond scientific visualization: Mapping information, Panel Proceedings, SIGGRAPH '90, pp 6–1 – 6–23.
14 "Visualizing the data," Algebra 2, senior authors Leiva MA and Brown RG, Cox interview; McDougal Littell A Houghton Mifflin Co., Evanston Boston Dallas, 1997, pp 255, 234–236, 271
15 Cox D (1994) Glitzy or grungy graphics: Choose your audience, AAAS '94 Program and Abstracts, for panel "Science, Lies, and Videotapes", San Francisco, CA, pp 93–94
16 Cox D (1988) Renaissance teams and scientific visualization: A convergence of art and science, Collaboration in Computer Graphics Education, SIGGRAPH '88 Educator's Workshop Proceedings, pp 81–104
17 Cox D (1987) Interactive computer-assisted RGB editor (ICARE), Proceedings for the 7th Annual Small Computers in the Arts Symposium, pp 40–45
18 Cox D (1991) Interdisciplinary collaboration case study in computer graphics education: "Venus and Milo," ACM SIGGRAPH Computer Graphics, Vol. 25, #3, pp 185–190

In Search of the Right Abstraction: The Synergy Between Art, Science, and Information Technology in the Modeling of Natural Phenomena

Przemyslaw Prusinkiewicz

Abstract

The creation of models of nature is the main objective of natural sciences. Without abstraction, however, models would be as complicated as reality itself; they would mimic nature without helping us to understand it. Identifying the essential features of the phenomena being described is therefore a crucial element of model construction. Unfortunately, an emphasis on objective, measurable characteristics, as promoted by current scientific practices, may lead in the wrong direction. An easily measurable characteristic may turn out to be irrelevant; on the other hand, a feature that eludes precise definition or measurement may be of central importance.

The paper illustrates this thesis by referring to the modeling of natural forms and patterns (in particular, plants) using the formalism of Lindenmayer systems combined with computer graphics visualizations. In this domain both precise botanical data and artistic observations play an important role. This synergy gives a new perspective to the centuries-old question of the relationship between science and art in describing the world around us.

1. Introduction

The tree shown in Fig. 1 does not really exist. The image is the result of a computer simulation based on a mathematical model of a horse chestnut tree [15]. The model captures the branching pattern inherent in this tree architecture, and includes a mechanism that modifies branching in response to local light conditions. Branching is most vigorous in abundant light, and decreases if the amount of light reaching a branch is reduced. Branches that are almost entirely in shade become a liability to the tree and are shed.

Is it a good model of reality? What does it mean to say a model is good, and how can we verify it? What is the role of simulation and visualization in model studies of the natural world? The purpose of this essay is to address these questions using plant modeling as an example.

2. What is a Good Model?

One intuitive criterion of a model's quality is its faithfulness, or the degree to

Fig. 1. A tree model with branches competing for access to light. From[15]

which it approximates reality. By itself, however, this criterion is insufficient, as illustrated by the following story from Suárez Miranda (1658), presented by Jorge Luis Borges under the title *On rigor in science*[3], and quoted by Umberto Eco in his essay *On the impossibility of creating a map of the empire in the scale of one to one*[7].

> ... in that Empire, the Cartographer's art achieved such a degree of perfection that the Map of a single Province occupied an entire City, and the map of the Empire, an entire Province. In time, these vast Maps were no longer sufficient. The Guild of Cartographers created a Map of the Empire, which perfectly coincided with the Empire itself. But Succeeding Generations, with diminished interest in the Study of Cartography, believed that this immense Map was of no use ...

The moral is that simplifications, or abstractions, are a necessary part of the modeling of nature. Without them, the models would mimic reality instead of explaining it.

The question is, what is the right abstraction? In areas of science with a long history of mathematical models – physics in particular – standard abstractions have been developed over the centuries and are widely accepted. For example, in classical mechanics one refers to masses and forces obeying Newton's laws of dynamics. Where the Newtonian approximation fails, different models provided by quantum mechanics and the theory of relativity are used. In life sciences, however, the problem of choosing the right level of abstraction is more difficult. For instance, the same forest will be considered differently by a botanist who studies the development of individual trees, an ecologist interested in the interactions between them, and an artist intent on capturing the beauty of the place. The choice of the right level of abstraction is a highly practical issue, which must be addressed each time a model is built[9, 19].

The question of choosing a good model when many models exist was considered in detail by Brian Gaines[8]. His point of departure is summarized by the following conversation:

> Imagine that you and I are each given the same sample of behavior and asked to model it from the same class of models. 'My model is a better approximation,' I say. 'Ah,' you reply, 'but mine is a far *simpler* model.'

Using the relationship between complexity and accuracy as the focal notion of his theory, Gaines considers a model admissible, if "any other model that gives a better approximation in accounting for the behavior is also more complex"[8]. Relaxing this definition, we may say that a good model offers a simple (maybe even the simplest possible) explanation of the observed phenomenon at the desired level of abstraction, and thus represents a favorable tradeoff between complexity and accuracy.

Thus, the ultimate goal of modeling nature is to construct simple yet faithful models of reality. This point has been succinctly summarized by Herbert Simon[21]:

> The central task of a natural science is to make the wonderful commonplace: to show that complexity, correctly viewed, is only a mask for simplicity; to find pattern hidden in apparent chaos.

According to this quotation, our perception of the complexity of nature reflects primarily the difficulty in recognizing (learning, inferring) nature's principles and regularities. Once they have been understood, faithful yet simple models can often be found.

3. Harnessing Model Complexity: An Example from Botany

How can we measure the complexity of a model? One approach, proposed by Kolmogorov (see [12]) is to express it as the length of the shortest description of a model. Strikingly compact, and therefore simple, models of nature can be found in the domain of botany. Below we consider an example to illustrate the point that simple models can indeed capture complex structures.

The key to identifying the regularities that underly plant forms is to look at them as dynamic systems that develop over time. In other words, in order to understand plant structures, we focus on the processes of development from which these structures emerge. This approach has a long tradition in biology, as presented by Thompson D'Arcy[22]:

> ... organic form itself is found, mathematically speaking, to be a function of time ... We might call the form of an organism an *event in space-time*, and not merely a *configuration in space*.

The processes of plant development can be conveniently expressed using the formalism of L-systems, introduced in 1968 by Aristid Lindenmayer[13]. An L-system consists of a set of *rewriting rules*, or *productions*, which capture the behavior of individual plant components (modules) over predetermined time intervals. For example, Fig. 2 shows a model of a compound leaf expressed using only two rules. The first rule states that an apex (i.e. a terminal branch segment) yields a branching structure consisting of two internodes, the apex continuing the main axis, and two lateral apices. The second rule states that, over the same time interval, the internodes will elongate by a factor of two. The bottom part of Fig. 2 shows a developmental sequence generated using these two rules. In each step, all apices and internodes are subject to their respective rules, applied in parallel. An intricate branching structure of a compound leaf results.

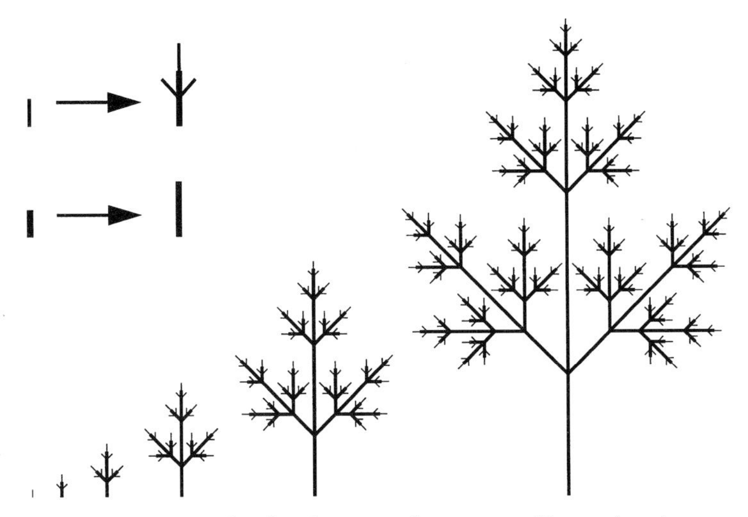

Fig. 2. The productions and a developmental sequence illustrating the operation of compound leaf model. The apices are shown as thin lines, the internodes as thick lines. From[17]

In the above example, the fate of each plant component was determined at the time of its creation. This corresponds to the biological notion of the control of development by lineage[14]. In nature, plants also employ more complex control mechanisms. These include endogenous interaction, in which control information flows through the growing structure (for example, in the form of hormones, nutrients, or water), and exogenous interaction, in which information flows through the space in which the plant grows[2,16]. With proper extensions[15,18], L-systems can capture both types of information flow, as illustrated by the tree model shown in Fig. 1. Although that model is no longer expressed by two simple rules, it is still remarkably compact (of the order of one hundred lines of L-system code).

4. Model Validation Through Simulations and Visualizations

Compactness offers a useful measure of model complexity. Let us now return to the other component of model quality, its faithfulness. How can we tell whether a model is a faithful representation of nature? To answer this question, let us consider the traditional process of constructing a scientific theory, which, for the purpose of this discussion, can be equated with the construction of a model (Fig. 3). As described by John Kemeny[11], this process begins with the observation of facts. The facts serve as the basis for constructing (inducing) a mathematical model, which is used to deduce predictions concerning the reality. An agreement between these predictions and new observations contributes to our confidence in the model.

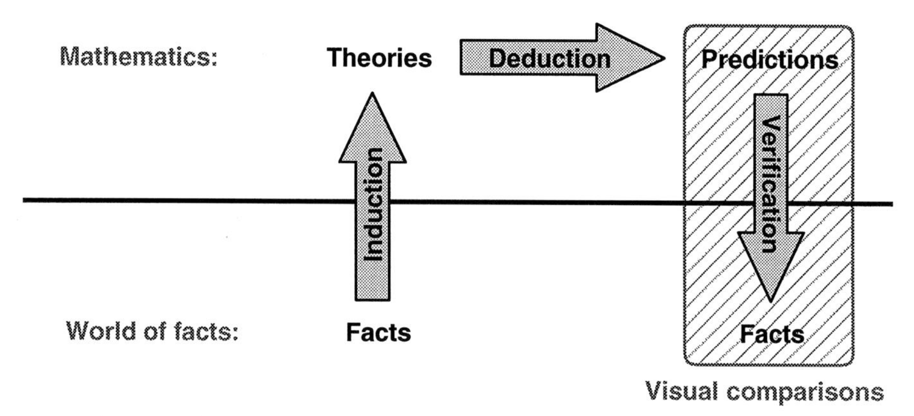

Fig. 3. Elements of the process of scientific discovery. Adapted from[11]

In the case of models expressed using L-systems, facts are the forms and developmental sequences of plants observed in nature. The models postulate mechanisms that may control plant development, and the predictions are the developmental sequences and forms resulting from the simulations. Comparison of the simulation results with the observations of reality is based primarily on visual inspection. We claim that Fig. 1 represents a faithful approximation or reality, because the generated structures looks similar to a real horse chestnut tree.

Is it acceptable to rely on visual comparisons while constructing and evaluating models of nature for scientific purposes? On the surface, this may seem highly subjective, qualitative, and unscientific. Nevertheless, given the state-of-the-art in characterizing arbitrary forms (for example, developed within the field of computer vision), there is not much more we can do. Satisfactory sets of parameters have been proposed for only relatively simple shapes. For example, we can measure how close an observed rounded shape is to a circle or a sphere[4], but we do not have comparable measures for complex shapes, such as trees. This is why visual comparisons play an important role at present.

This observation raises a question regarding the role of realism in the visual depiction of models. Is it necessary to represent the results of simulations realistically for visual comparison purposes? Maybe more schematic representations would be sufficient? The following example demonstrates that realism is important indeed. Fig. 4 looks like an unorganized collection of lines, distributed more densely at the top than at the bottom. Nevertheless, it represents the same tree as shown in Fig. 1, except that all the branches have been drawn using lines of

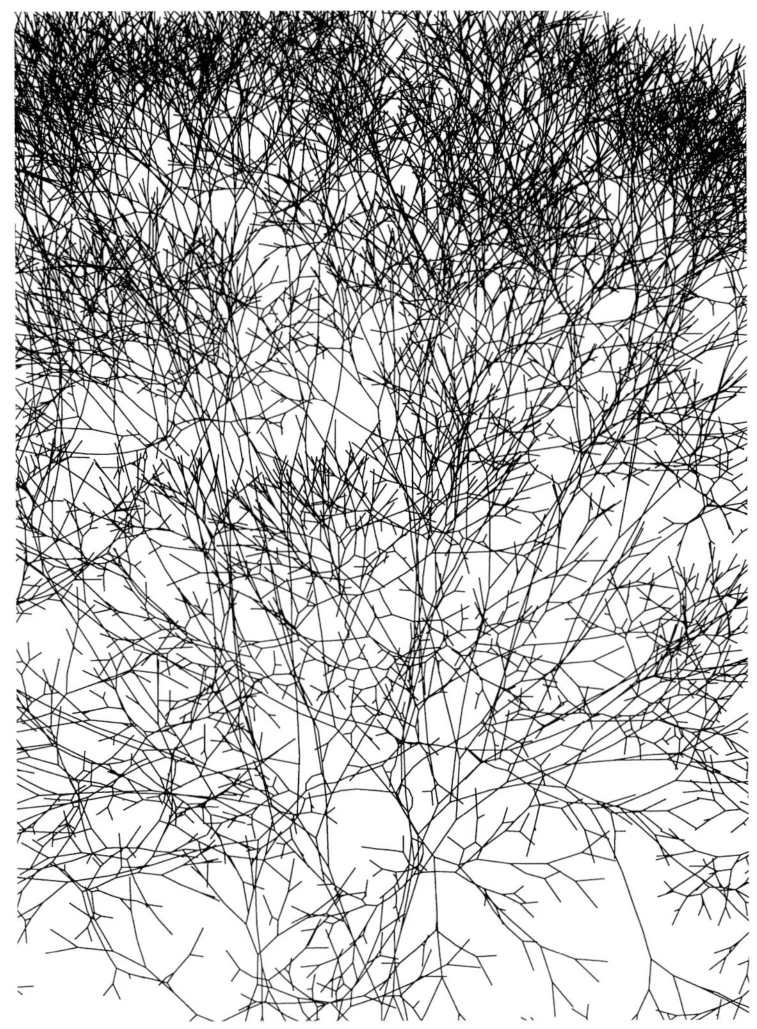

Fig. 4. Tree model from Fig. 1., rendered using lines of constant width

constant, equal width. It is difficult to abstract from the arbitrary artifacts in the visual presentation of models, even if they are conceptually small. While carefully chosen simplifications may be useful, realistic representation of the modeled objects is the safest choice for visual comparison purposes.

5. Conclusions

We have sketched the logical path that brings answers to the fundamental questions of the modeling of nature, stated in Section 1. These answers highlight the fact that modeling is an interdisciplinary activity which combines a knowledge of the modeled phenomena, the computer science methodology for creating the simulations, and the craft of visualizing the models for validation purposes. In its highest form, this craft becomes an art, fulfilling the role phrased by Pablo Picasso:

> Art is the lie that helps us see the truth.

The synergy between art and science also has another facet: that of creating images of nature for aesthetic reasons. In this case, an understanding of nature serves as the basis for an artistic process. In the domain of botany, the mutually enriching relationship between art and science has a particularly long and well documented history, which can be traced back to Leonardo da Vinci[6]. An interesting case in point is the comparison of two texts on trees from the beginning of the twentieth century: Rex Vicat Cole's *Artistic Anatomy of Trees*[5] and H. Marshall Ward's botanical treatise *Trees. Form and habit*[23]. In spite of the differences in objectives and points of view, these books contain many strikingly similar observations and illustrations.

Computer science complements the synergy between art and science in two major ways. First, computer simulations and visualizations are a means for better understanding scientific models. This application of computer science has been emphasized by Alan Kay[20]:

> The strength of our culture over the past hundred years has been our ability to take on multiple points of view. That's what simulations allow you to do.

Second, computer science contributes to the process of scientific discovery by providing a focus on control mechanisms and the flow of control information in nature. This perspective, well manifested in L-system plant models, has its roots in cybernetics[1] and was recently stated by Gruska and Jürgensen[10]:

> 'Computer science' should be considered as a science with aims similar to those of physics. The information processing world is as rich and as important as physical world for mankind.

Acknowledgments

I would like to thank Ryohei Nakatsu and Christa Sommerer, whose kind invitation for me to attend the ART-SCIENCE-ATR Symposium lie at the origin of this essay. I am also indebted to Lynn Mercer for most useful comments on the manuscript, and to Jules Bloomenthal for many inspiring discussions on art, science, and computer graphics (as well as reference[5]). The underlying scientific work has been sponsored by grants from the Natural Sciences and Engineering Research Council of Canada, and a Killam Resident Fellowship.

References

1 Apter MJ (1966) Cybernetics and development. International Series of Monographs in Pure and Applied Biology/Zoology Division Vol. 29. Pergamon Press, Oxford

2 Bell AD (1986) The simulation of branching patterns in modular organisms. Philos. Trans. Royal Society London, Ser. B, 313: 143–169

3 Borges JL (1972) A universal history of infamy. Translated by N. T. di Giovanni. Dutton, New York

4 Boyer M, Stewart NF (1991) Modeling spaces for toleranced objects. The International Journal of Robotics Research, 10(5): 570–582

5 Cole RV (1915) The artistic anatomy of trees. Seeley, Service, London (republished by Dover, New York, 1965)

6 Richter JP (ed) (1970) The notebooks of Leonardo da Vinci, compiled and edited from the original manuscripts by Jean Paul Richter, in two volumes. Dover Publications, New York

7 Eco U (1994) How to travel with a salmon and other essays. Harcourt Brace, New York

8 Gaines B (1997) System identification, approximation, and complexity. Int. J. of General Systems, 3: 145–177

9 Godin C, Guédon Y, Costes E, Caraglio Y (1997) Measuring and analysing plants with the AMAP software. In: Michalewicz MT (eds) Plants to ecosystems. Advances in computational life sciences I. CSIRO Publishing, Melbourne

10 Gruska J, Jürgensen H (1990) Informatics: a fundamental science and methodology for the sciences (emerging from Computer Science and maturing). Manuscript, Department of Informatics, Slovak Academy of Sciences, Bratislava, and Department of Computer Science, University of Western Ontario, London, Ontario

11 Kemeny J (1959) A philosopher looks at science. Van Nostrand, Princeton

12 Li M, Vitányi P (1993) An introduction to Kolmogorov complexity and its applications. Springer Berlin Heidelberg New York Tokyo

13 Lindenmayer A (1968) Mathematical models for cellular interaction in development, Parts I and II. Journal of Theoretical Biology, 18: 280–315

14 Lindenmayer A (1982) Developmental algorithms: Lineage versus interactive control mechanisms. In Subtelny S, Green PB (eds) Developmental order: Its origin and regulation. Alan R. Liss, New York, pp. 219–245

15 Měch R, Prusinkiewicz P (1996) Visual models of plants interacting with their environment. Proceedings of SIGGRAPH '96 (New Orleans, Louisiana, August 4–9, 1996) ACM SIGGRAPH, New York, pp. 397–410.

16 Prusinkiewicz P (1994) Visual models of morphogenesis. Artificial Life, 1(1/2): 61–74

17 Prusinkiewicz P, Hammel M, Hanan J, Měch R (1997). Visual models of plant development. In: Rozenberg G, Salomaa A (eds) Handbook of formal languages. Springer Berlin Heidelberg New York Tokyo (to appear)

18 Prusinkiewicz P, James M, Měch R (1994) Synthetic topiary. Proceedings of SIGGRAPH '94 (Orlando, Florida, July 24–29, 1994). ACM SIGGRAPH, New York, pp. 351–358

19 Remphrey WR, Prusinkiewicz P (1997). Quantification and modelling of tree architecture. In: Michalewicz MT (eds) Plants to ecosystems. Advances in computational life sciences I. CSIRO Publishing, Melbourne
20 Ryan B (1991) Dynabook revisited with Alan Kay. Byte, 16(2): 203–208
21 Simon HA (1969) The sciences of the artificial. MIT, Cambridge
22 D'Arcy Thompson (1952) On growth and form. Cambridge University Press, Cambridge
23 Ward HM (1909) Trees. Volume V: Form and habit. Cambridge University Press, Cambridge

Artificial Life for Computer Animation

Demetri Terzopoulos

Introduction

Computer animation, an endeavor firmly situated at the confluence of art and science, may be characterized as the art of animating mathematically formulated models of objects by computer. The state of the art of computer animation in feature films is represented by the production *"Jurassic Park"* (Universal Pictures, 1993), in which many of the dinosaur special effects were computer-generated, and by the first feature film made entirely using computers, *"Toy Story"* (Walt Disney Productions/Pixar, 1995). These superb technical milestones make it easy for audiences enjoying them to overlook the fact that the animate dinosaur and toy characters are by no means autonomous, living creatures. In fact, the characters are simple "graphical puppets" which are laboriously animated by human animators. These artists are, in essence, highly skilled "puppeteers". They are capable of making conventional graphics models (that typically specify little more than the 3D shape and appearance of a character) move and act in a lifelike manner. Computer animators are well versed in techniques, such as keyframe interpolation and motion capture, which help mitigate the labor of their craft and they also rely on computers to render moving images of their animated subjects.

This article describes a new approach to the realistic computer animation of living creatures that differs decidedly from the state of the art reviewed above. We confront the scientific challenge of developing *functional* computer graphics models of animals. The modeling of animals is a major theme in the burgeoning field of *Artificial Life*[1]. Drawing upon concepts from artificial life, we have developed sophisticated graphics models that are *self-animating,* thus potentially relieving computer animators from the drudgery of animating conventional models. Like natural animals, our *artificial animals* are endowed with functional bodies and brains. Artificial animals are *virtual autonomous agents* that inhabit simulated physical worlds.

In particular, we have created realistic artificial animals in the form of fishes. Imagine a virtual marine world inhabited by a variety of self-animating fishes (Fig. 1). In the presence of underwater currents, the fishes employ their muscles and fins to swim gracefully around immobile obstacles and among moving aquatic plants and other fishes. They autonomously explore their dynamic world in search of food. Large, hungry predator fishes stalk smaller prey fishes in the deceptively peaceful habitat. Prey fishes swim around contentedly, until the sight of preda-

(a)

(b) (c)

Fig. 1. Artificial fishes in their virtual marine world as it appears to an underwater observer. (a) The three reddish fish are engaged in mating behavior while the others are foraging among seaweeds. (b) A school of fish appears in the distance. (c) A predator shark stalking prey

tors compels them to take evasive action. When a dangerous predator appears in the distance, similar species of prey form schools to improve their chances of survival. As the predator nears a school, the prey scatter in terror. A chase ensues in which the predator selects victims and consumes them until satiated. Some species of fishes seem untroubled by predators. They find comfortable niches and feed on floating plankton when hungry. Driven by healthy libidos, they perform fascinating courtship rituals to secure mates.

Animation as Artificial Life Cinematography

Applying our artificial life approach to computer animation, we have produced two computer-animated short subjects that have been screened internationally before large audiences [2,3]. The creative process underlying these animations, which are essentially mini-documentaries about the virtual marine world of artificial fishes, has the following distinguishing feature: Rather than being a graphical model puppeteer, the computer animator has a job more analogous to that of an underwater nature cinematographer. Immersed in the virtual marine world, the animator strategically positions one or more virtual cameras to capture interesting "film footage" of the behaviors of artificial fishes. The footage is edited, narrated, and assembled to produce the final documentary. Thus, the creative process is in certain ways similar to the one associated with the fascinating genre of marine life documentaries produced by the *Cousteau Society* or by the *National Geographic Society*.

In the remainder of the article, I will review the artificial fish models that have made possible our *artificial life cinematography* approach to computer animation. The bottom-up, compositional modeling methodology, in which we model the form, appearance, and basic physics of the animal and its habitat, the animal's means of locomotion, its perceptual awareness of its world, its behavior, and its ability to learn, has been described in detail elsewhere [4–7]. Our methodology is generally applicable to the modeling of animals other than fishes for use in computer animation.

Capturing Form and Appearance

We want our artificial fishes to capture the form and appearance of a variety of natural fishes with considerable visual fidelity. To this end, digitized photographs of real fish, such as the images shown in Fig. 2a, are converted into three-dimensional spline surface body models (Fig. 2b) with the help of interactive image analysis tools, and the image texture is mapped onto the surfaces to produce the final textured geometric display models of the fishes (Fig. 2c).

In conventional computer animation, the animator would have to rely on skill and would labor with standard techniques such as rigid-body transformation,

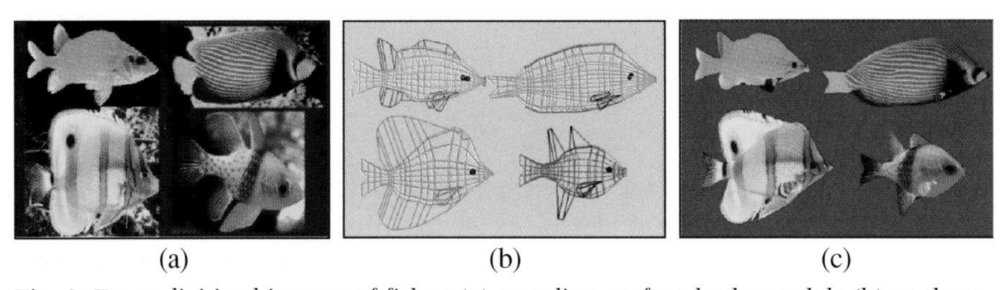

(a) (b) (c)

Fig. 2. From digitized images of fishes (a) to spline surface body models (b) to three-dimensional, textured fish display models (c)

free-form deformation, and keyframe animation in attempting to satisfy the challenge of making the animal display models – be they dinosaurs, toys, or fishes – locomote and behave in a realistic manner. Instead, we proceed to develop a more complete, functional model of the animal that can automatically animate itself with remarkable realism.

Functional Modeling

The artificial fish is an autonomous agent with a realistic deformable body actuated by internal muscles, with eyes, and with a brain that includes behavior, perception, and motor centers. Fig. 3 presents an overview of the functional model.

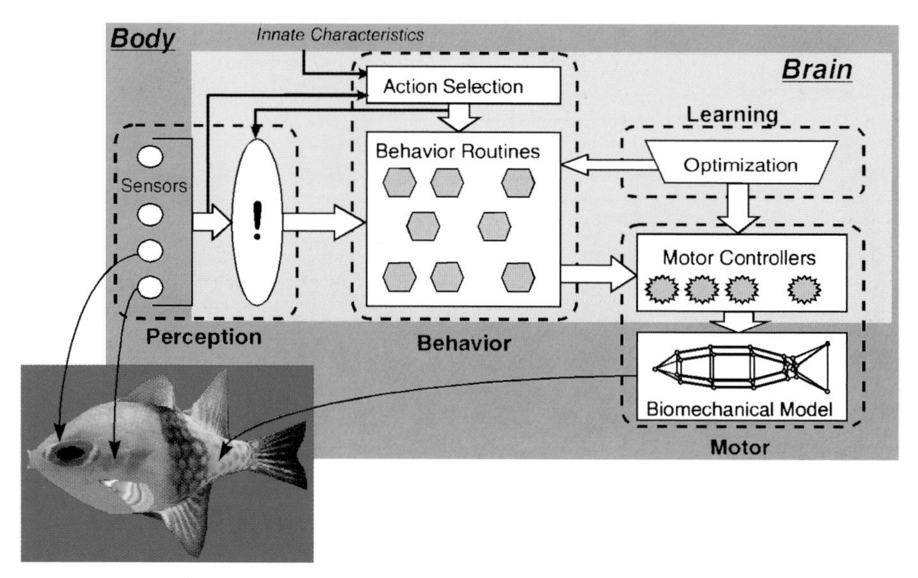

Fig. 3. Functional artificial fish model. The piscine body harbors a biomechanical model and a brain with motor, perception, behavior, and learning centers. (The perceptual attention module is marked "!")

Through controlled muscle actions, artificial fishes are able to swim through simulated water in accordance with hydrodynamics. Their functional fins enable them to locomote, maintain balance, and maneuver in the water. Thus the artificial fish captures not just the 3D form and appearance of real fishes, but also the basic physics of these animals in their environment. Though rudimentary compared to their natural piscine counterparts, the brains of artificial fishes are nonetheless able to learn some basic motor functions and carry out perceptually guided motor tasks. In accordance with their perceptual awareness of the virtual world, their minds arbitrate a repertoire of piscine behaviors, including collision avoidance, foraging, preying, schooling, and mating. The following sections present the four main functional components of the artificial fish.

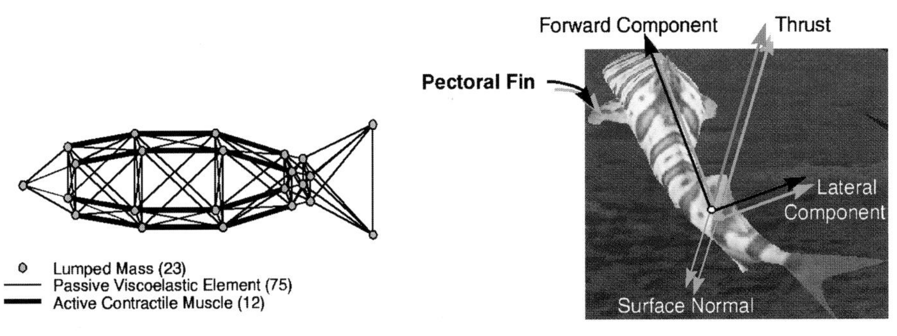

Fig. 4. Biomechanical fish model (a). Nodes denote lumped masses. Lines indicate uniaxial elastic elements. Bold lines indicate muscle elements. Hydrodynamic locomotion (b). With the tail swinging to left, the thrust on any point on the body acts opposite to the surface normal at the point. The forward thrust component propels fish through the simulated water

Biomechanics and Locomotion

The motor system of the artificial fish (refer to Fig. 3) comprises a piscine biomechanical model, including muscle actuators and a set of motor controllers. Fig. 4a illustrates the mechanical body model which produces realistic piscine locomotion using only 23 lumped masses and 87 elastic elements. These mechanical components, whose dimensions and physical parameters are modified to model different fishes, are interconnected to maintain the structural integrity of the body as it flexes due to the action of its 12 contractile muscles.

The artificial fish locomotes like real fishes do, by autonomously contracting its muscles. As the body flexes it displaces virtual water which induces local reaction forces normal to the body. These hydrodynamic forces generate thrust that propels the fish forward (Fig. 4b). The dynamics of the biomechanical model are governed by a system of coupled second-order ordinary differential equations driven by the hydrodynamic forces. A numerical simulator continually integrates these equations of motion forward through time. The biomechanical model achieves a good compromise between realism and computational efficiency, permitting the simulation of fish locomotion in real-time (on an R10000 class Silicon Graphics workstation).

The model is sufficiently rich to enable the design of motor controllers by gleaning information from the fish biomechanics literature. The motor controllers (see Fig. 3) coordinate muscle actions to carry out specific motor functions, such as swimming forward, turning left and right, ascending and descending. They translate natural control parameters such as the forward speed or angle of the turn into detailed muscle actions that execute the function. The artificial fish is neutrally buoyant in the virtual water. Its two pectoral fins enable it to navigate freely in its three dimensional world by pitching, rolling, and yawing its body. Additional motor controllers coordinate the fin actions.

Perception

Artificial fishes are aware of their world through sensory perception. Their perception system relies on a set of on-board virtual sensors to gather sensory information about the dynamic environment. As Fig. 5 illustrates, it is necessary to model not only the abilities but also the limitations of animal perception systems in order to achieve natural sensorimotor behaviors.

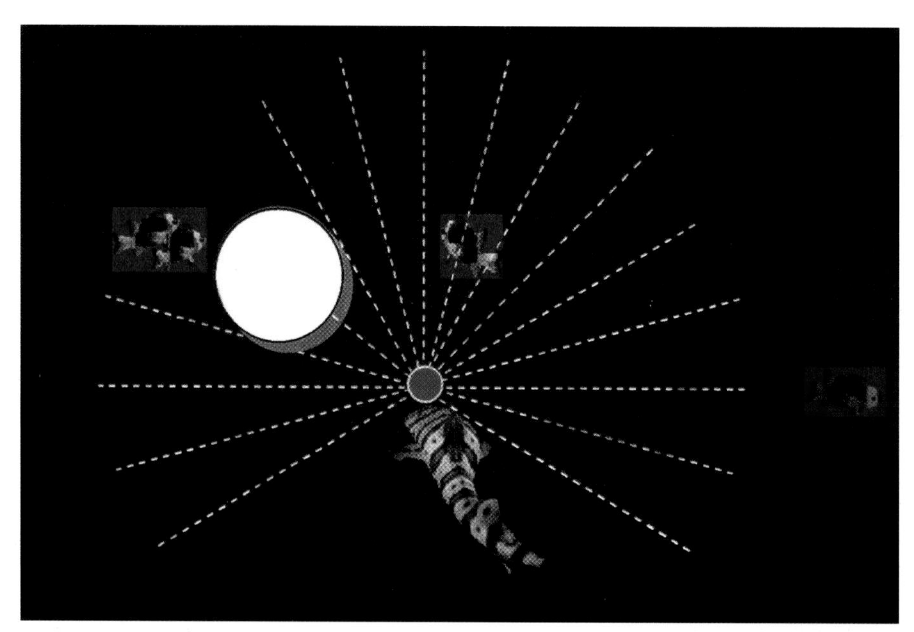

Fig. 5. Artificial fishes perceive objects within a limited field view if objects are close enough and not occluded by other opaque objects (only the fish towards the left is visible to the fish at the center)

The perception center of the artificial fish brain includes a perceptual attention mechanism (indicated by "!" in Fig. 3) which enables the artificial fish to sense the world in a task-specific way, hence filtering out sensory information superfluous to its current behavioral needs. For example, while foraging, the artificial fish attends to sensory information about nearby food sources.

Behavior

A set of prespecified innate characteristics determine whether the fish is male or female, predator or prey, etc. The behavior center of the artificial fish's brain mediates between its perception system and its motor system (see Fig. 3). The behavior system runs continuously within the simulation loop. An intention generator, the fish's cognitive faculty, harnesses the dynamics of the perception-action cycle. The intention generator is responsible for the goal-directed behavior of the artificial fish in its dynamic world.

At each simulation time step, the intention generator takes into account the innate characteristics of the fish, its mental state as represented by *hunger*, *fear*, and *libido* mental variables, and the incoming stream of sensory information, to generate dynamic goals for the artificial fish, such as to avoid an obstacle, to hunt and feed on prey, or to court a potential mate. The intention generator ensures that goals have some persistence by exploiting a single-item memory. Persistence supports sustained behaviors such as foraging, schooling, and mating. The intention generator also controls the perceptual attention mechanism. At every simulation time step, it activates behavior routines that attend to sensory information and compute the appropriate motor control parameters to carry the fish one step closer to fulfilling its current intention. The behavioral repertoire of the artificial fish includes primitive, reflexive behavior routines, such as obstacle avoidance, as well as more sophisticated motivational behavior routines, such as schooling and mating, whose activation depends on the mental state.

Learning
The learning center of its mind (refer to Fig. 3) enables the artificial fish to learn how to locomote through practice and sensory reinforcement. Through optimization, the motor learning algorithms discover muscle controllers that produce efficient locomotion. Muscle contractions that produce forward movements are "remembered". These partial successes then form the basis for subsequent improvements in swimming technique. Their brain's learning center also enables the artificial fishes to train themselves to accomplish higher level sensorimotor tasks, such as maneuvering to reach a visible target or learning more complex motor skills (see[6,7] for the details).

As an application of our learning algorithms, in reference[6] we describe, among other artificial animals, the construction of an artificial dolphin that has learned to perform the sort of stunts that elicit applause at marine theme parks like *SeaWorld*. For example, we can present the artificial dolphin with the task of leaping out of the water and it is capable of learning how to accomplish the task. In particular, the artificial dolphin discovers that it must build up momentum to perform the leap by thrusting upwards from a starting point deep in the pool. Fig. 6a shows a snapshot as the dolphin exits the water. The dolphin can also learn to perform acrobatic tricks while in the air. Fig. 6b shows it using its nose to bounce a large beach-ball off a support. The dolphin can learn to control the angular momentum of its body while exiting the water and during ballistic flight so that it can perform aerial rolls and somersaults. Fig. 6a shows it in the midst of a somersault in which it has just bounced the ball with its tail instead of its nose. Fig. 6d shows the dolphin right after a dramatic splashdown. By learning to control its body in complex ways, the dolphin can amuse and entertain the animator and the audience.

Conclusion
In this article, I have demonstrated how the science of artificial life can contribute to the art of computer animation. In particular, I described a virtual marine world inhabited by artificial life forms that emulate the appearance, motion, and beha-

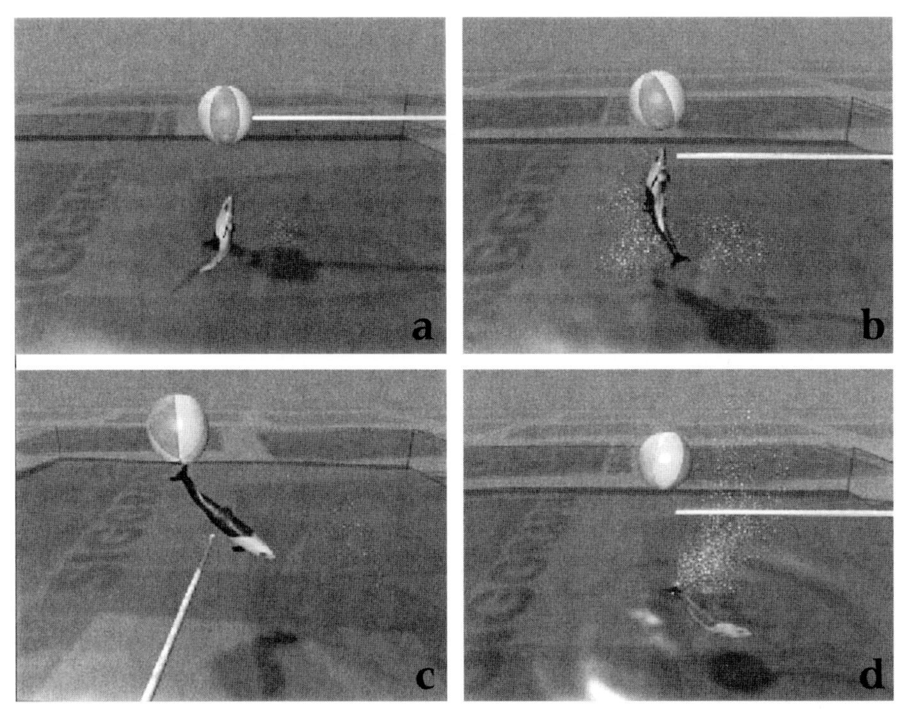

Fig. 6. *"Sea World"* leaping tricks learned by the artificial dolphin. The dolphin leaps out of the water (a) and bounces the ball either with its nose (b) or with its tail (c) before reentering the water with a splash (d)

vior of marine animals in their natural habitats. Each artificial fish is an autonomous agent in a simulated physical world. It has (i) a three-dimensional body with internal muscle actuators and functional fins that deforms and locomotes in accordance with biomechanic and hydrodynamic principles, (ii) sensors, including eyes that can perceive the environment, and (iii) a brain with motor, perception, behavior, and learning centers. Artificial fishes exhibit a repertoire of piscine behaviors that rely on their perceptual awareness of their dynamic habitat. Furthermore, they can learn to locomote through practice and sensory reinforcement.

Our artificial life approach to computer graphics modeling, as exemplified by artificial fishes, is also applicable to the realistic modeling of other animals. This unconventional approach has enabled us to produce realistic computer animation of natural environments in which the animator plays a role akin to that of a nature cinematographer. In our animated productions, the detailed motions of the artificial fishes emulate the complexity and unpredictability of movement of their natural counterparts, which enhances the visual beauty of the animations.

Acknowledgements
I thank my students Xiaoyuan Tu, Radek Grzeszczuk, and Tamer Rabie for their out-

standing contributions to the research reviewed herein. The research has been made possible by grants from the Natural Sciences and Engineering Research Council of Canada and the support of the Canadian Institute for Advanced Research.

References

1 For an engaging survey of the Artificial Life field, see, e.g., Levy S (1992) "Artificial Life" Pantheon. Journals such as "Artificial Life" and "Adaptive Behavior" (MIT Press) document the state of the art
2 Tu X, Grzeszczuk R, Terzopoulos D (1995) A National Geo-Graphics Society Special: The Undersea World of Jack Cousto. Computer animation premiered at the ACM SIGGRAPH '95 Electronic Theater, Los Angeles, CA
3 Tu X, Terzopoulos D, Fiume E (1993) Go Fish! Computer animation in ACM SIGGRAPH Video Review Issue 91: SIGGRAPH '93 Electronic Theater
4 Terzopoulos D, Tu X, Grzeszczuk R (1994)Artificial fishes: Autonomous locomotion, perception, behavior, and learning in a simulated physical world." Artificial Life, 1(4): 327–351
5 Tu X, Terzopoulos D (1995) Artificial fishes: Physics, locomotion, perception, behavior. In: Computer Graphics. Proceedings, Annual Conference Series, Proc. SIGGRAPH '94, Orlando, FL, pp 43–50
6 Grzeszczuk R, Terzopoulos D (1995) Automated learning of muscle actuated locomotion through control abstraction. In: Computer Graphics. Proceedings, Annual Conference Series, Proc. SIGGRAPH '95, ACM SIGGRAPH, Los Angeles, CA
7 Terzopoulos D, Rabie TF, Grzeszczuk R (1996) Perception and learning in artificial animals. In: Artificial Life V: Proc. Fifth International Conference on the Synthesis and Simulation of Living Systems, Nara, Japan, pp 313–320

3. Artificial Life

Evolution as Artist

Thomas S. Ray

Abstract

Evolution is a creative process, which acting independently, has produced living forms of great beauty and complexity. Today, artists and engineers are beginning to work together with evolution. In the future, it may be possible for artists to work in collaboration with evolution to produce works of art whose beauty and complexity approach that of organic life. A variety of styles of creative collaboration between humans and evolution are discussed.

The human arts: painting, sculpture, music, cinema, etc., are creative expressions. Similarly, life forms: plants, animals and humans themselves, are also creative expressions. In the case of the arts, the creative force is the human imagination, whereas in the case of organisms, the creative force is evolution.

Human artists express themselves in many media: oil paint, clay, stone, music, cinema, etc. Until recently, the creative expressions of evolution have been known only from a single medium: organic chemistry. Life on Earth is the creative expression of evolution by natural selection working in the medium of carbon chemistry.

The creative products of evolution include the human body and mind, the cheetah running down its prey, the mahogany tree, the humming bird pollinating a flower. These living works of art exceed in beauty, and depth of structure and process, anything produced by the best of human artists. In fact, human artists themselves (and therefore human art) are products of evolution.

When we observe the creative products of evolution with our naked senses, we see only a single level, for example, the visual surface of an orchid. On this surface we can see great beauty, richness, subtlety and complexity of structure. However the richly organized structure of living systems is much deeper than what meets the eyes.

Evolution has organized the form and process of matter and energy on Earth from the molecular level up through the level of the ecosystem, spanning a range of twelve orders of magnitude of scale. At each level of this range, evolution has created complex forms and processes, with each level being built hierarchically from those below and forming the basis of those above.

At every level, the forms are as rich and beautiful as what we see with the naked eye, though at most levels the aesthetic is less conventional. Our unaided vision allows us to directly observe living structures of sizes ranging from small individual organisms, up through landscapes which can encompass entire ecosystems. These images are in the domain of conventional aesthetics to which we are well tuned.

As we move outside of the range of what we can normally visualize, we encounter forms in living systems with a similar quality of richness, subtlety and complexity, but which require an unconventional aesthetic to appreciate. For example the forms of ecosystems and metabolic pathways are based on the flows of matter and energy through these systems. These are rich organic forms, but they can not be directly visualized. Probably most scientists who study these systems develop an aesthetic appreciation for them. However, this is a rare aesthetic, based on a specialized education in the life sciences.

1. Evolution in Other Media

Life on Earth is the product of evolution in the medium of carbon chemistry. However, in theory, the process of evolution is neither limited to occurring on the Earth, nor in carbon chemistry. Just as it may occur on other planets, it may also operate in other media, such as the medium of digital computation. Just as human artists are able to express their creativity in a variety of media, so can evolution.

In recent decades, there have been a wide variety of implementations of evolution in the digital medium. While these have varied greatly in the degree to which the selection is natural or artificial, it is essentially the same creative evolutionary process that has expressed itself so richly in the organic medium on Earth.

Contemplating the richness of organic evolution raises the question of how richly the same process can express itself in the digital medium. In the experiments conducted to date, we have seen dramatic transformations in the replicators, however, we have not seen the emergence of hierarchically organized complex structures comparable to the products of organic evolution.

2. Collaborating with Evolution

How might we best work with evolution? How can we collaborate with evolution in a way that enhances our own creative expressions? How can we help evolution to reach its full potential for creative expression? There are many approaches to working with evolution. The different forms of collaboration vary in the degree to which we control the evolutionary process by artificial selection, or free it to operate by natural selection, and in the degree to which the genetic language pre-determines the form of the replicators.

At one extreme, evolution is reduced to the role of technician or builder, filling in the details, by optimization, of a design completely specified in advance by the human. At the other extreme, evolution is free to generate and refine its own designs, with the human only providing the raw materials for evolution to work with. In between these extremes there exist more equitable collaborations in which evolution generates creative alternatives and the human provides aesthetic (or other selective) guidance.

2.1 Genetic Algorithms

The most highly constrained form of collaboration is typified by the field of "genetic algorithms" (GA)[1,3]. Generally, genetic algorithms use bit strings to encode the solutions to some engineering problem. By mutating and crossing over bit strings in a large population, repeatedly evaluating the "fitness" of the solutions, and preferentially replicating the most fit solutions, the genetic algorithm can search for optimal solutions.

During the design of the GA, the manner in which the bit string encodes the solution space is determined. Normally this involves a fixed length bit string, with successive portions of the string assigned to the representation of predetermined quantities, such as the coefficients in an equation. Thus the form of the solution is determined in advance, and does not take part in the evolutionary process.

Although researchers in the GA field freely use the phrase "natural selection" to describe their evolutionary process, the GA absolutely does not use natural selection. It uses artificial selection. The designer of the GA writes a "fitness function" algorithm, which determines which members of the population of bit strings will be favored through replication.

It is also worth noting that the strings in a GA do not self-replicate. They are copied by the simulation system, after evaluation by the fitness function. They may be copied precisely, or with some "mutations" in the form of bit flips, or in many cases, only a portion of the bit string is copied, in combination with a complementary portion from another favored bit string.

The GA represents one extreme in the spectrum of control: total control. The GA software totally controls: the form of the solution, the fitness function, the nature of the genetic operators, and the method of replication. This approach makes minimal use of evolution's creative potential.

2.2 Genetic Programming

In a more recent development, known as "genetic programming" (GP), the solutions are defined by trees of lisp-like expressions. The genetic operations of mutation and cross-over can operate at any node of the tree. In the case of cross-over, a node is chosen at random in (typically) two different trees. The nodes, and all of their higher branches and leaves are simply swapped between the trees.

In this method, the form of the solution does not have to be defined in advance, and so can also evolve. Although the GP exhibits a relatively free-form solution space, it shares with the GA a total control in the nature of the "fitness function" and the process of replication. By permitting evolution to determine the form of the solution, GP allows a more creative use of the process of evolution, and has been applied to a wide array of problems, notably by John Koza[4].

3. Karl Sims – Virtual Creatures

Sims used a similarly flexible genetic system to specify solutions to the problem of being a "creature" built of collections of blocks, linked by flexible joints powered by "muscles" controlled by circuits and possibly including sensors[10,11] (Fig. 1). Sims embedded these block creatures in simulations of real physics, such as in water,

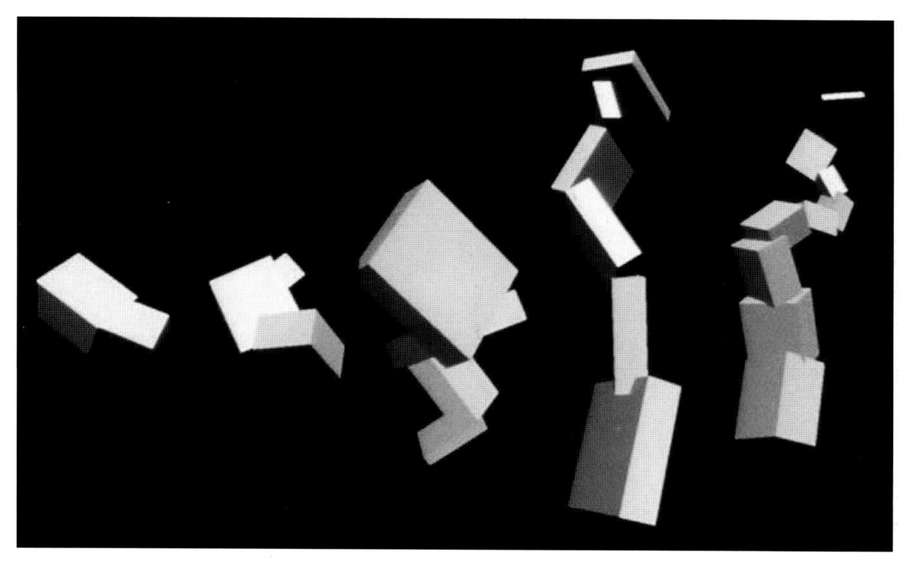

Fig. 1. Virtual Creatures. © Karl Sims at Thinking Machines

or on a surface. They were then selected for ability at a variety of tasks, such as swimming, swimming after a light source, moving across a surface, and jumping on a surface.

These experiments produced a bewildering and fascinating array of creatures. Some were familiar such as the swimming "snake" and walking "crab" forms. Others were effective at their tasks, but accomplished them with completely unfamiliar patterns of movement and form. This wild proliferation of forms recalls that which occurred among the animals of the Cambrian explosion (see below), when many more experiments were attempted than have survived to be familiar to us today.

In essence, this work is comparable to a GA or GP, but the genotypes are based on graphs rather than strings or tree structures. Like any optimization, the fitness function is predefined (e.g., maximum swimming velocity). However, the objective of the work was not to find an optimal solution, but rather a diversity of solutions for each fitness function. The design of the system was such that a stunning diversity of solutions was possible and did emerge.

Rather than presenting the results as an optimization curve, showing how performance (e.g., maximum velocity) improves over evolutionary time, Sims presented the results in the style of natural history. He presented the form and behavior of a diversity of evolved individuals.

A loosening of the fitness function took place in an experiment where the objective was to possess a block (Fig. 2). Possession of the block involved a competition between two co-evolving populations, tested in individual pairs. Therefore, the form of the competing creature was an evolving part of the fitness landscape. In this case, a large component of the fitness function was outside of the specifi-

Fig. 2. Virtual Creatures. © Karl Sims at Thinking Machines

cation of the system. Thus this system exhibits a significant component of natural selection, and allows greater creativity on the part of evolution.

2.4 Aesthetic Selection

An intermediate level of evolutionary creativity was achieved by Sims, in his "genetic images" work. This work involved an unusually balanced collaboration between the human and evolution. Sims used the genetic programming method in combination with selection based on the aesthetic criteria of the user to evolve abstract images[9] (Fig. 3). The fitness function is provided by the aesthetically based selections of the human user in each generation of images. These selection criteria are whimsical and change in each generation as the genetic operators generate new arrays of choices. In this collaboration, evolution is constantly suggesting new designs, while the human artist provides aesthetic judgement and guidance to select among evolution's suggestions.

3. Evolution as Artist

In order to maximally exploit the creative potential of evolution, it is necessary for the human collaborator to give up most of their control over the process. The

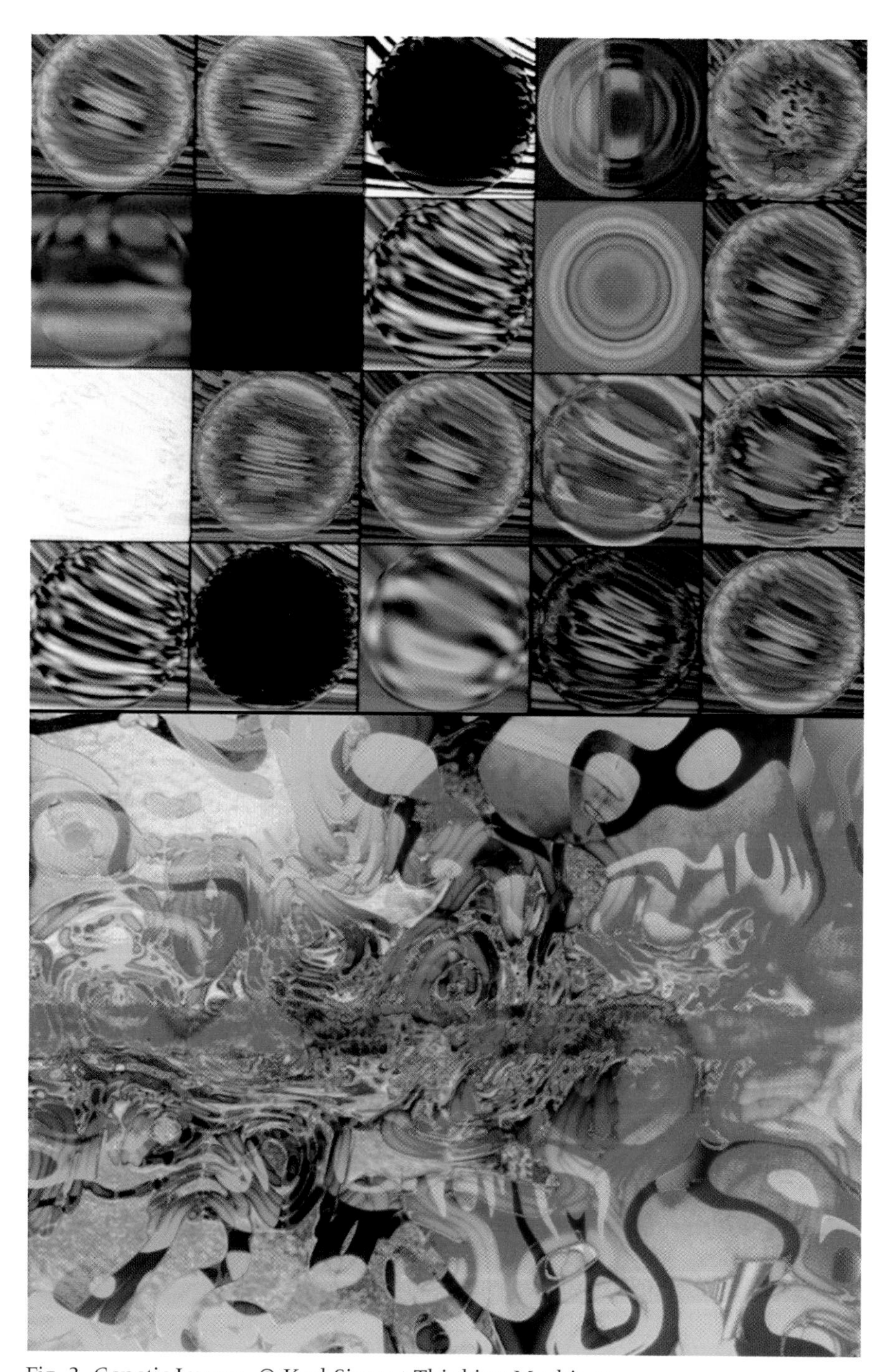

Fig. 3. Genetic Images. © Karl Sims at Thinking Machines

human only sets up the environment for evolution to operate in, provides it with raw materials, and then watches as evolution expresses its creativity. This means that the human does not provide any guidance to evolution, and thus can not necessarily expect evolution to produce a useful product. But it is under these conditions that evolution has the maximal freedom to express its own creativity.

We do not know yet, if we can ever expect evolution in the digital medium to express a level of creativity comparable to what we have seen in the organic medium. However, it is likely that evolution can only reach its full creative potential, in any medium, when it is free to operate entirely by natural selection, in the context of an ecological community of co-evolving replicators.

Before developing this idea further, I want to discuss some relevant aspects of the history of organic evolution on Earth. The fossil record indicates that evolution shows brief but dramatic bursts of creativity, against a background consisting primarily of variations on and elaboration of themes which originate during the creative bursts. Earth's most creative evolutionary transitions were reviewed recently[5]. Some of the major innovations noted were: origin of chromosomes, origin of eukaryotes, origin of sex, origin of multi-cellular organisms, and origin of social groups.

Of these major transitions, perhaps the most dramatic, and best known, was the rapid origin and diversification of large multi-cellular organisms from microscopic single celled ancestors, in what has come to be known as the Cambrian explosion of diversity. It has understandably been called evolution's "big bang", when there was a dramatic inflation of complexity of organisms, and species diversified rapidly into an ecological void[2].

In trying to bring out the full potential of evolution in the digital medium, we should attempt to create the conditions under which analogous fundamental transitions can occur. Otherwise, we are likely to always be operating at the level of variations on existing themes, without any fundamentally new innovations.

4. Evolving Complexity

In the traditional approach to working with evolution, as exemplified by plant and animal breeding and the fields of genetic algorithms and genetic programming, humans guide the evolution of a population of replicators by deciding which members of the population are allowed to reproduce. The breeder provides a "fitness function" or a "selection criteria".

Breeding manages evolution within the species, producing variations in the forms of existing species. However, evolution is also capable of generating new species. Even more significantly, evolution is capable of causing an explosive increase in the complexity of replicators, through many orders of magnitude of complexity. The Cambrian explosion may have generated a complexity increase of eight orders of magnitude in a span of three million years. Harnessing these enormously more creative properties of evolution requires an approach completely different from traditional breeding.

We know how to apply artificial selection to convert poor quality wild corn into high-yield corn. However we do not know how to breed algae into corn.

There are two bases to this inability: 1) if all we know is algae, we could not envision corn; 2) even if we know all about corn, we do not know how to guide the evolution of algae along the route to corn. Our experience with managing evolution consists of guiding the evolution of species through variations on existing themes. It does not consist of managing the generation of the themes themselves.

As a thought experiment, imagine being present in the moments before the Cambrian explosion on Earth, and that your only experience with life was familiarity with bacteria, algae, protozoa and viruses. If you had no prior knowledge, you could not envision the mahogany trees and giraffes that were to come. We can't even imagine what the possibilities are, much less know how to reach those possibilities if we could conceive of them.

Imagine that a team of Earth biologists had arrived at a planet at the moment of the initiation of its Cambrian explosion of diversity. Suppose that these biologists came with a list of useful organisms (rice, corn, pigs, etc.), and a complete description of each. Could those biologists intervene in the evolutionary process to hasten the production of any of those organisms from their single celled ancestors? Not only is that unlikely, but any attempts to intervene in the process are likely to inhibit the diversification and complexity increase itself.

If the silk moth never existed, but we somehow came up with the idea of silk, it would be futile to attempt to guide the evolution of any existing creature to produce silk. It is much more productive to survey the bounty of organisms already generated by evolution with an eye to spotting uses for existing organisms.

5. Digital Biodiversity Reserve

In an attempt to create conditions under which evolution might express itself in the digital medium with a creativity analogous to what we have seen in the organic Cambrian explosion, I have proposed the creation of a "biodiversity reserve for digital organisms", a kind of wildlife reserve in cyberspace. The network installation of Tierra[6–8] creates a new web within the internet that is inoculated with digital organisms which are allowed to evolve freely through natural selection.

This web will have a complex topology of interconnections, reflecting the topology of the internet within which it is embedded. In addition, there will be complex patterns of "energy availability" (availability of CPU cycles). Consider that each node on the net tends to experience a daily cycle of activity, reflecting the habits of the user who works at that node. The availability of CPU time to the Tierra process will mirror the activity of the user, as Tierra will get only the cycles not required by the user for other processes. Statistically, there will tend to be more "energy" available for the digital organisms at night, when the users are sleeping.

There will be selective pressures for digital organisms to maintain themselves on nodes with a high availability of energy, by sensing and responding to temporal and spatial patterns of resources (such as CPU "energy"). This might involve daily migrations around the planet, keeping on the dark side. In short, the digital organisms must be able to intelligently navigate the net in response to dynamically changing circumstances.

In addition to responding to conditions on the net itself, digital organisms evolving in this environment will have to deal with the presence of other organisms. If one node stood out above all the rest, as the most energy rich, it would not be appropriate for all organisms to attempt to migrate there. They wouldn't all fit, and if they could they would have to divide the CPU resource too thinly. Thus there will be selection for social behavior, flocking or anti-flocking behavior. The organisms must find a way of distributing themselves on the net in a way that makes good use of resources.

The complex topology of the network, and the heterogeneous and changing patterns of energy availability should favor the selection of behavior more complex than simple reproduction. It is hoped that this will launch evolution in the direction of more complexity. Once this trajectory has begun, the interactions among the increasingly sophisticated organisms themselves should lead to further complexity increases.

In the original single node Tierra, most of the evolution observed involved the adaptation of organisms to other organisms (parasitism, immunity, social behavior, etc.)[6] (Fig. 4). It is this kind of dynamics that can lead to an auto-catalytic increase in complexity and diversity in an evolving ecological system. The complexity of the physical system in which evolution is embedded does not have to lead the complexity of the living system.

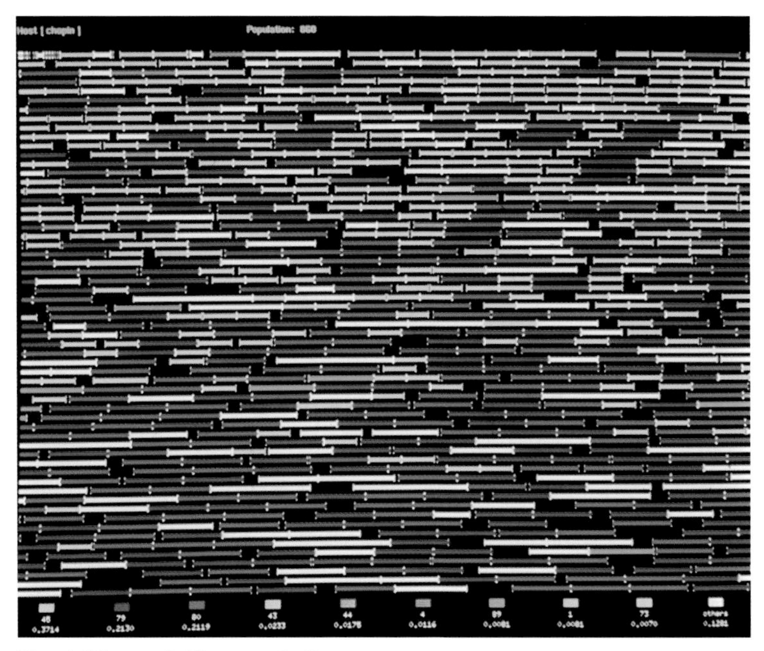

Fig. 4. Tierra. © Thomas S. Ray

6. Reaping the Harvest

The strategy being advocated here is to let natural selection do most of the work of directing evolution and producing complex digital organisms. These will be "wild", living free in the digital biodiversity reserve. In order to reap the rewards, and create useful products, we will need to harvest these organisms, and in some cases domesticate the wild digital organisms, much as our ancestors domesticated dogs and corn thousands of years ago. Some of the useful products from organic evolution are: rice, corn, wheat, carrots, beef cattle, dairy cattle, pigs, chickens, dogs, cats, guppies, cotton, mahogany, tobacco, mink, sheep, silk moths, yeast, alligators, and penicillin mold. All of these products were derived from organisms that were spontaneously generated within an ecosystem of organisms evolving freely by natural selection. Some of these are used essentially in their wild form, and may be harvested directly from nature, but many are farmed and domesticated.

Humans have been managing the evolution of other species for tens of thousands of years, through the domestication of plants and animals. It forms the basis of the agriculture which underpins our civilizations. We manage evolution through "breeding", the application of artificial selection to captive populations.

The process of harvesting the products of digital evolution must begin with observation. Digital naturalists must explore the digital "jungle", observing the natural history, ecology, evolution, behavior, physiology, morphology, and other aspects of the biology of the organisms of the digital ecosystem, like modern day tropical biologists exploring our organic jungles.

However, occasionally, these digital biologists may spot an interesting information process for which they see an application. At this point, some individuals can be captured and brought into laboratories for closer study, and to farms for breeding. Sometimes, breeding may be used in combination with genetic engineering (insertion of hand written code, or code transferred from other digital organisms). The objective will be to enhance the performance of the process for which there is an application, while diminishing undesirable wild behavior. Some digital organisms will domesticate better than others, as is true for organic organisms (alligators don't domesticate, yet we can still ranch them for their hides).

It seems obvious that organisms evolving in the network-based biodiversity reserve will develop adaptations for effective navigation of the net. Yet at this point we surely can not conceive of where evolution in the digital domain will lead, so we must remain observant, imaginative in our interpretations of their capabilities, and open to new application possibilities.

7. New Aesthetics

The products of a completely free evolution in the digital medium are not likely to be inherently visual or auditory in nature. Thus they would not be recognized as conventional artistic creations. However, just as life scientists have been able to develop an aesthetic appreciation for the non-visualized forms of living nature, we may also be able to develop new aesthetics to appreciate the beauty of the products of digital evolution. If evolution can express its creativity in the digital medium, with a richness comparable to what it has expressed in the organic medium, then the products of digital evolution should have a comparable richness of form and process, which will surely be beautiful to behold.

References

1 Goldberg DE (1989) Genetic algorithms in search, optimization, and machine learning. Addison-Wesley, Reading, MA
2 Gould SJ (1989) Wonderful life. W. W. Norton, pp. 347.
3 Holland JH (1975) Adaptation in natural and artificial systems: an introductory analysis with applications to biology, control, and artificial intelligence. Univ. of Michigan Press, Ann Arbor
4 Koza JR (1992) Genetic programming, on the programming of computers by means of natural selection. MIT, Cambridge, MA
5 Maynard Smith J, Szathmáry E (1995) The major transitions in evolution. Freeman, Oxford
6 Ray TS (1991) An approach to the synthesis of life. In: Langton C, Taylor C, Farmer JD, Rasmussen S (eds) Artificial Life II, Santa Fe Institute Studies in the Sciences of Complexity, vol. X. Addison-Wesley, Redwood City, CA, pp 371–408
7 Ray TS (1994) An evolutionary approach to synthetic biology: Zen and the art of creating life. Artificial Life 1(1/2): 195–226. Reprinted in: Langton CG (eds) Artificial Life, an overview. MIT, Cambridge, MA
8 Ray TS (1995) A proposal to create a network-wide biodiversity reserve for digital organisms. ATR Technical Report TR-H-133. Available by ftp at tierra.slhs.udel.edu as /tierra/doc/reserves.tex
9 Sims K (1991) Artificial evolution for computer graphics, Computer Graphics (Siggraph '91 proceedings), Vol. 25, No.4, pp 319–328
10 Sims K (1994) Evolving virtual creatures, Computer Graphics (Siggraph '94) Annual Conference Proceedings, pp 15–22
11 Sims K (1994) Evolving 3D morphology and behavior by competition. In: Brooks R, Maes P (eds) Artificial Life IV Proceedings. MIT, Cambridge, MA, pp 28–39

Artificial Life under Tension – A Lesson in Epistemological Fabulation

Louis Bec

1. Background

Artificial life[1] is a construct that accommodates a tensorial space.[2]

Thus it is the result of marked tension between the living and the technologically created near-living.

Its techno-ecosystemic niche presents as a 'potential' for chimerization.[3] It is subject to a tension existing between life defined as an intrinsic property of matter and life redefined as a technological simulation device.

This tension describes a distinctive trajectory in the overall relationship between the arts and the sciences. Thus it opens up entirely new fields of exploration and plays a part in the current reconfiguration of knowledge and forms of expression. This trajectory traverses the scientific, artistic and technological domains, in all their diversity, evolutions and mutations. Via the multiplicity of interactions thus generated it gives rise to offset 'epistemological and esthetic tensions'. It generates valuable indicators allowing for a truly fine-tuned reading of the multiple relationships now hybridizing the sciences of the living, artistic experimentation and the biotechnologies.

This tensorial space, with its remarkable energy content, causes artificial life to suffer distortions between its phylogenetic and ontogenetic boundaries, as well as distensions and retractions between the biomimetic phases and the techno-zoosemiotic possibilities that map its future.

It mingles and brings into conflict certain epistemological and methodological

---

[1] "Artificial Life is the study of manmade systems exhibiting behavior characteristic of natural living systems.
Via attempting simulation of 'live analogous' behaviors on computers and other artificial media, it complements the traditional approach of the biological sciences, whose mode of functioning is the analysis of living creatures. In its extension of the empirical foundations of biology beyond the carbon chains of earth-based organisms, artificial life can contribute to theoretical biology by situating life as we know it in the broader context of life as it might be."
Chris Langton, Artificial Life, Santa Fe, 1988

[2] Set of numbers making up a system used to represent the tensions in a space or a solid.

[3] Chimerization: the process of creating zoological chimeras using parts of different organisms.

dissensions[4] existing between the postulates of theoretical biology and the bio-computerized approach, and between robotics and the instrumentology of simulations of the living. It acts as a mutagenic agent within technology-linked artistic approaches.

The areas of representation and modeling of the living are explored by elasticity deformation. The newer areas of interactivity, digital information and networks, the programming of processes relating to movement, real time, virtual space and man/animal/machine interfaces, are explicitly designated as fundamental activities for experimental creation.

In fact, artificial life is inhibited by the animate schema of the living.

Its role is that of an autonomous behavioral agent, in techno-sensorial[5] interaction with the fluctuating environment of that knowledge out of which, patiently and via its own learning capacity, it develops its behavior and its inventive adaptation.

The artistic and scientific convergences and divergences artificial life testifies to are based on a primeval tremor.

An imperceptible tremor of the living, a vibration going back to time immemorial. By giving rise to a logical proliferation wave, it compels recognition of the 'pro-creation' of techno-biodiversity[6] as a fundamental mode of human expression.

2. Modeling

Modeling is the conceptual and operational tensor of Artificial Life.

Generally speaking, in the artistic as well as the scientific areas the aim of modeling is to produce a median object situated between the data and the model-maker.

As the bearer of variable, manipulable parameters this median object allows for the processing of emotional, imaginative information in an artistic setting[7] in the same way as for logical, rational information in a scientific one.

The artificial life artefact programmatically compacts quantitative and syntactic data and qualitative and semantic data, in order to produce a 'meta-model' by chimerization or new 'artificial metabolic potentialities' by modeling.

---

[4] Theoretical biology seems to take no interest in behavioral creations effected in substrata other than organic terrestrial molecules.
Simulation raises the problem of the nature of the phenomenon reproduced, for it cannot be claimed that the model really captures the essence of the phenomenon. What interest is there in a project that claims to study *life as it might be while being* validated by life as we know it ?

[5] Techno-sensorial: refers to the oversizing of sensoriality by the use of technical prostheses.

[6] Techno-biodiversity: increase in the diversity of living organisms via to the constructions of artefacts.

[7] When painting, for example, is set the task of using matter to reproduce a phenomenon or capture an emotion, it becomes a variable model allowing for constant parameter manipulation.

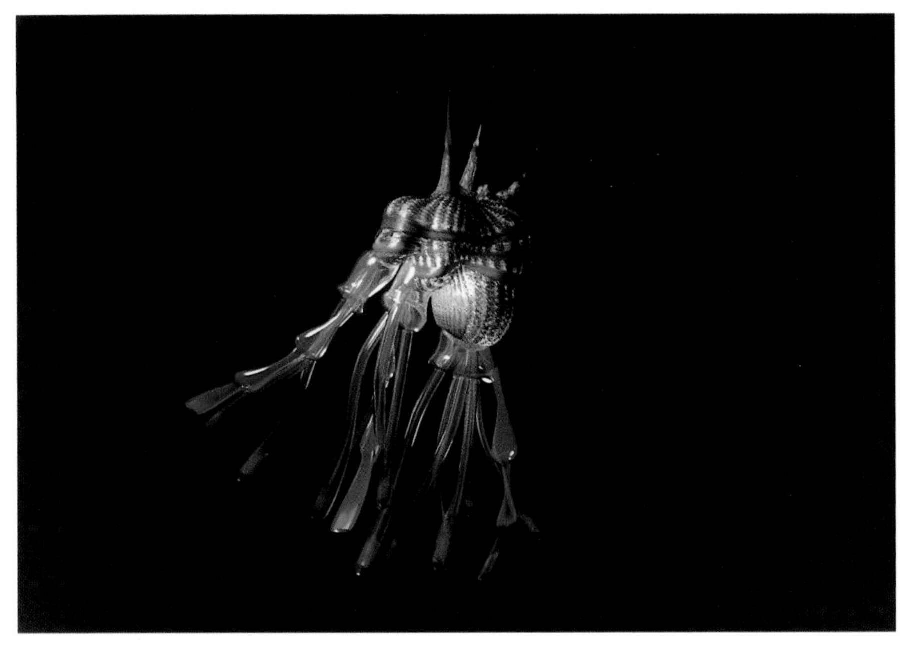

Fig. 1. Triklope klor. © Louis Bec

> Modeling is thus in itself the construction of a variable tension between a phenomenon to be simulated and a device that simulates.

Artificial life modelings give concrete expression to choices made between analytic modeling strategies and systemic or heuristic ones. And so, when modeling tends to reduce the distance between the phenomenon under study and its artefact, analogical modeling results. In this case, the model will obey the principles of connection or similarity. The choice of parameters and of their experimental variabilities will give rise to formal, behavioral and cognitive representations or will describe the successive infrastructure or process states duplicated.

The result is an esthetics of mimesis.[8]

In material terms, this takes the form of a realism that consists of imitating life as it is. Digital technologies are fascinated by the representation of reality as stabilisation of tension. Thus the majority of Artificial life modelings belong implicitly and explicitly to the logic of analogy, resemblance and biomimetics. More interesting, on the other hand, are those new forms of esthetics that seem to be developing via autonomy or the principles of programmatic delegation. They consist in giving 'flexibility' to the artificial life artefact, via a non-programmed freedom capable of generating unexpected 'emotional' events; in this they reproduce the behavior of an independent or interdependent living thing.

[8] The notion of a model is from the outset implicit in the definition of *mimesis*, of imitation: to state (with Aristotle) that art *imitates* nature is equivalent to saying that art takes nature as its model, in terms of its object, even more so of its functioning and maybe of its underlying principle.

Dynamic modeling of complex systems takes place when the system integrates the model into its own tensions as an active system agent. The agent modifies the behavior of the system and as a consequence modifies itself.

Artificial life modeling is no longer the representation, effected from the outside, of a living organization. It presents as a self-organizing, autonomous, tensorialized entity.

This modeling takes the form of interactive devices, evolutional morphogeneses, simulation of universes that vary according to data implementation or are seized via different capture modes.

> We can regard the esthetics of Artificial Life as an emergence pointing up the active substance of the systems constructed.

It becomes the mode of expression of the totality of the tense or slack variable relationships between the system's set of components, the behaviors that develop within it, and the knowledge and technologies used.

The symptoms of this esthetics lie in the successive states of an 'elasto-stylistic' proliferation of artificial events that leaves traces or trajectories of variable speed and duration in virtual or alternative universes.

It is an emersion of a sunken revelation. It is an esthetics of refloatation[9] which exhibits to the living intelligible determiners of the living.

The living itself means to grasp them via its curious faculty to identify via self-identification.

It must be capable of establishing a cunning, malleable dialogue between the living and the near-living via a shared algorithmic language; but also of authenticating the indicators shown so as to assign them to one of the catagories of a living extended to include the artificial.

Every living entity must refloat its livingness so as to recognize what the living is. In precisely this way it also artificializes itself extendably.

3. *Biotelematics*

Positivist epistemology has weighed heavily on the proponents of artificial life.

For a long time it prevented the carrying out of a modeling of life as it might be.

Anchored in a knowledge object and not in a knowledge projection, biomimetic modeling[10] has had the effect of putting the emphasis on biphase connection modeling and not on tensile coherence modeling.

---

[9] A sunken hulk gradually becomes covered with elements of the living, at the same time as it is transformed by rust and the rotting of wood.
Its technical structure becomes the substratum and the skeleton for a multitude of living species. The effects of the undersea environment, the adaptive and behavioral processes of the organisms and the physical, mechanical and chemical changes undergone by the boat bring into play the chimerization between the naval technology and the benthic living world.
The intrication here is so close-knit that the refloated boat can be regarded as a living artefact.

[10] Biomimetic modeling drew on three types of functioning of the living: autoregulation, autopreservation, autoreproduction. Via cellular automata, genetic algorithms, collective intelligence modeling, dynamic morphogeneses and behavioral robotics there emerges a simulation copied directly from the living.

Fig. 2. Melaskunodousse. © Louis Bec

We know that artificial life develops according to the principles of extensive viability and of proliferation and exploration strategies. The conditions for bringing about technological biodiversity, for the creation of unknown forms of life and post-biological virtual worlds, can thus only come about in a new, truly and totally artificial space.

> Cyberspace is the artificial place where the tension of digital communication is at work.
> A mass of information circulates in the form of electric and light impulses in an artificial space that has become a planetary electromagnetic field.

Artificial life intends to infiltrate proliferating digital biotelematic organisms into this environment and colonize the communication networks.[11]

This new environment possesses a material layer made up of cables, Hertzian circuits, optical fibers, satellites, computers, microchips and so on. In this environment the machines are commutators linked to machines and the silicon microchips are interconnected.

Considered as a virtual provider of the living, this environment could contribute to the emergence of unknown binary life-forms; these would make up, so to speak, a hypotechnozoology[12] capable of appearing in all the network's visualization terminals.

---

[11] See the work of Thomas S. Ray and his Tierra project.

[12] Hypotechnozoology: a zoology emerging from the underside of positivist zoology via technological modeling.

The environment can model their behavior and point up other types of adaptive abilities.

It can establish a technozoosemiotics of coding and cryptography and generate duplicative activities via technical autoreplication artefacts. It can facilitate the incubation of distributed intelligence colonies capable of multiplying in all the network's dimensions and complexity. It can secrete dynamic non-linear morphogeneses appropriate to the communication protocols.

This set of operations necessarily entails structural, textual, textural and behavioral mutations via encoding/decoding, cryptography, and data compression/decompression/depression.

4. Technozoosemiotics

Artificial life is now emerging from a data-processing space. It generates an information surplus.

> It is a technozoosemiotic tension produced by the connection of the living with the near-living.
> It directs the semaphoric infrastructure[13] of an artificial life cooped up in its conventional modeling of the strata of the living.[14]
> Henceforth it compels recognition as a transducer/translator, between signal and sign.

Its new field of activity involves setting up overall communication between the living organisms of the biomass, via technological interfaces, while also establishing relationships with other 'natural or artificial' forms of intelligence implanted in other media or evolving in other spaces.

This displacement points up the extent, not of the classical difference between the living and the artificial, but of another tension, that of an 'inter-code' showing that along with denoted, functional signals there also exist, between the living and the near-living, diffuse, connoted, punctual signals that interconnect them.

Artificial Life is situated at the intersection of the multiple exchanges linking all the components of the biomass and the natural and technological environment. The information surplus resulting from these reciprocal activities has to be processed by the interfaces of a 'biologically living/technological/data processing/instrumentological' complex.

Thus artificial life as technozoosemiotics elaborates conversational agents and inserts transduction and transcoding spaces between kinesthetic, paralinguistic systems and grammaticalized, discretized forms of language.

Artificial life as technozoosemiotics is situated at the crossroads of semiotics, ethology, the cognitive sciences, technology, computer science and artistic practise; it is an integral part of zoosemiotics, which studies the signals elaborated by living species for inter- or extra-specific communication.

---

[13] Technosemaphorics: a technological sign-bearing device.

[14] The different phases that have characterized the living: animation, mechanism and organization.

Working permanently and diffusely, this transversal interspecific communication can be regarded as one of the fantastical issues involved in artistic creation.

Art is haunted by the living, by animality, by its representations and modelings.

Body art, the introduction of the living into the plastic arts[15], current interactive artificial life devices – all of which redefine the living as an expressive substance in its own right – have not got to the bottom of two still unanswered questions:

1. Is it not the case that, at the very heart of artistic endeavor, there has long been the demiurgic ambition to create the living via multiple simulations?

Do cloning, genetic engineering and the creation of transgenic animals open the way to terratological art?

2. Communication with other species has yet to work. But do not domestication, zootechnics, animal cognition studies and the search for paralanguages, and extraterrestrial biology testify to an irresistible need for interchange with the other belonging to a species originating elsewhere?

> This is why it is not surprising to see artistic approaches appearing which get out from under the usual psychological pressures and 'situate man's creative activities in the lineage of those of animals and other living organisms'.[16]

These approaches draw on technological advances, concepts and modeling in the cognitive sciences, animal and human ethology in its search for animal protoculture as the roots of culture, the origin of language, voice-recognition, the emergence of higher symbolic functions and so on.

Certain lines of artistic research have become genuinely involved in interspecific animal communication.[17]

In their combining of expressive modalities such as image, sound and text – but also behavior, movement and gesture – are transdisciplinary artistic, scientific and technological approaches also leading to the making of tomorrow's tools for widespread interspecific communication?

Had we been more attentive, maybe we would have spotted signs of what was to come. We probably need to go back to Beuys's famous tête-a-bête with a coyote in America.

This performance has to be seen as a premonitory act, a still unfocused attempt to point out that the future of a unequivocal, totally new and richly promising artistic activity lies in the logosystemic[18] establishment of a 'danced' relationship between all the components of the biomass.

---

[15] Kounellis was one of the first artists to exhibit living forms in the gallery.

[16] René Atlan, Créativité biologique et auto-création du sens ('Biological Creativity and Selfcreated Meaning'). Castella Albeuve, Switzerland, 1986

[17] Animal Art, Steirischer Herbst, Graz, 1987

[18] Logosystemics: a system made up of the elements of a discourse in their interaction.

Transition of Concept of Life in Art and Culture from Automata to Network

Machiko Kusahara

Introduction

Life has been always an important theme of arts. While what life means has been one of the major issues of biology as well as philosophy, artists have also dealt with the theme from different aspects.

The history of life science and the history of art seem to flow toward the same direction. Before Darwinism it was believed that all the creatures were designed by the Creator and remained the same since then. The hierarchy among the Creator and human beings, and other animals was absolute.

Major discoveries in life science that took place since 1950 have changed our concept of life. Molecular biology told us that life is a phenomenon that can be described in terms of physics, chemistry and information science; and that human beings are made of the same material and codes as other animals. Modern biology deals with notions such as gene, emergence, self; concepts which are more related to information and process rather than material and the goal. The fact that information is the key element that supports the diversity of life and the mechanism of evolution made a direct link between life science and computer.

And computer has also become a medium of art.

Artists can create virtual time-space with virtual life forms. Life as a theme of art can be now life itself, not only the way of life. With the advent of digital technology such as computer graphics and virtual reality, what used to be represented in an abstract manner can be visualized with convincing reality. Artists now seek for a different approach to life, with suggestions given from biology or other fields of science. But they are also using biological concepts from different approaches, developing interesting ideas that would help our culture to understand life with a richer context. Life can be a metaphor of art.

I would like to show how the concept of creating life has been developed in our history. Scientific discoveries have played an important role in establishing our knowledge which has become the basis of our way of thinking. But on the other hand science cannot develop without an appropriate background in our society. It might be through such socio-cultural development that both science and art develop toward the same direction, while interacting with each other, and that is the reason why certain convergence occurs between the two fields of human activity.

Because of that I will also describe how the same questions regarding our life are interpreted in a different culture – namely Japanese – with a different background. The description might look too detailed sometimes as I try to cover some points which might be hard to understand outside from the culture.

We should not be trapped in a single point of view. As history of science, art and technology is too often described and discussed only from occidental point of view, I hope such observation on Japanese media art would offer a hint for a further analysis on the meaning of science and technology in art.

Art Changes

Art experienced a big change both in concept and in method in the beginning of the century. Artists discovered film as a medium for representing illusions and imaginations. In fact, techniques such as animation and film tricks made it possible to create a virtual world within films[1]. Optical illusions became commonplace. The medium of art is not bound to material any more.

On the other hand DADA movement brought a question to the role of artists and the nature of artworks. The artists were fascinated by machinery both in real and metaphorical ways, and used the new media such as photography and film for art making.

Since then the paradigm of art has continued to change. The most important concept in terms of media art was brought by Fluxus and related art movements proposing a new paradigm on the relationship between artists and viewers, often using electronic medium for such purpose[2]. Viewers can be involved in the activity of art as a part of it. In other words, art would become an environment that viewers can step into and interact with. Meanwhile increasing public awareness on ecology brought up genres of arts such as environmental art, earthworks and land art. What is common in these is the denial of the traditional concept of art where an artwork is made solely through the creative activity of the artist and should remain untouched.

Now we have computer graphics to create illusions from numbers. Interactive technology and virtual reality are in the reach of artists. These tools fulfill the long term desire of art-making to construct a 3-dimensional imaginary world that we can interact with.

The nature of art, its material, and its relationship to our life have changed. This change did not occur by itself but developed along the way we see ourselves. Science played a major role in the transition. What we see now is not just the influence of the science on art but the convergence of science and art especially in the notion of life.

---

1 There is much to be told about the rich history of pre-cinema equipment and the applications. Such technology as magic lanterns was the new medium of that time and was used both in showing realistic images and illusions. It was even used for scientific education

2 Kusahara M, Sommerer C, Mignonneau L (1996) Art as living system. In: Kusahara M (ed) The state of interactive art. InterCommunication Vol. 7

Life and Media

The rise of 'Artificial Life' (A-Life) took place at the right moment when life science was no longer a specific area for biologists. It offers an arena to analyze and discuss the nature of life that yields to its abstraction and generalization. For artists, life has achieved a different role as a medium of art and communication through such abstraction .

Shakespeare was quite right when he said that we are made of the same material as dreams are made of[3]. Of course Shakespeare was not talking about bits and bytes. But if the essence of life consists in information, dreams and imagination can be simulated in the same manner as representation of the real world.

More recent topic in biology and medicine has drawn the wide attention of people including philosophers and artists. It is the research on the manner how a living organism distinguishes itself from others. Immunity is based on such recognition of "self". When a recognition mechanism malfunctions the body starts attacking itself. On the other hand an originally alien life such as a mitochondria resides in our cells without being excluded. Questions such as how a microscopic organism defines itself by making distinctions between the self and others, or what is the nature of "self" at the level of organisms without any sort of consciousness, bring about more general question on our body and mind. "Self", a traditional theme of art and literature now confronts a new paradigm.

Research on complexity has been attracting wide attention. The concept is likely to be the key issue in many fields. Our culture has been influenced greatly by biological concepts and facts such as evolution, gene, DNA, ecology, immunity (or the identification of self). Art is not an exception. But it is not necessarily a one-way influence.

Unnecessary to mention Charles Darwin; scientists who worked in the field of life science have had great influence on our culture. The points of view they develop through their research go beyond the limit of scientific papers. Researchers such as Jacques Monod, Francois Jacob, Richard Dawkins, Steven Gould, or the Japanese Takeshi Yoro are among them. What is the case of artists working with concepts in/from life science? How do such concepts influence their works, and what do they bring to art and culture? Do their works go beyond the world of art to influence science?

Objective View of Life

> How can the events in space and time which take place within the spatial boundary of a living organism be accounted for by physics and chemistry?... The obvious inability of present-day physics and chemistry to account for such events is no reason at all for doubting that they can be accounted for by those sciences. (E. Schrödinger)[4]

---

[3] "Tempest"

[4] Schrödinger E (1944) What is life? The physical aspect of the living cell, p 2

The argument that life can be accounted for by objective and logical events held by Schrödinger and Nils Bohr led molecular biology and biophysics.

In one aspect, this argument led the concept of life as information inspired by the discovery of the double helix by Watson and Crick. In another aspect, by statistically evaluating energy, which can be measured, it was possible to comprehensively describe the life phenomena from the micro to the macro scale. Statistical evaluation means that the temporal development of life phenomena is not determined at any instant. It signifies the complexity of the process.

Schrödinger is not the first to try to account for life events using physics principles. The first physicist who reduced all facets of life into physical elements to explain the universe was Aristotle. The representative thought in Greek explains that objects live with the existence of an anima[5], but will die once it is lost. Aristotle could not conceive that all organisms uniformly possess anima so that he noted that the amount of anima present in an organism depends on the type of organism and the state of the organism[6]; a view that the stages of life can be determined by the amount of anima. As he found the problem in the binary stance on the existence/absence of life, he tried to solve the problem by introducing the concept of the degree of life.

On the other hand, the ancient Chinese had separated the soul which governs the intellect and reasoning and the soul which gives an animal like behaviour in life. That there exists some middle ground if the balance or timing between the two is collapsed was the Chinese way of thinking. In "Peony Lantern", the ghost story of the Chinese classics, the dead (or can one call this the dead?) in this state cause problems. Even after the soul of the mind leaves the body, the soul of the body lingers in the flesh, purely bodily functions occur without the reasoning or common sense. This describes as two different souls governing, the different functions now ascribed to different parts of the brain. It is a practical interpretation which corresponds to situations where the human consciousness is lost but the body continues to "live".

The background of such idea would go back far to the elements we find in Asian culture. The same elements have brought different ways of thinking in terms of such concepts as self, originality, life and death, the status of human beings in our religion, etc. In general, vagueness of boundary is the underlying feature in Asian view of life compared to that in the western world[7].

---

[5] The word "animation" used also as in computer graphics motion pictures signifies that anima is endowed to something which was originally not living. In that sense, we are still using the Greek philosophy?

[6] Emmeche C (1994) The garden in the machine, p 25

[7] For example, in popular traditional Buddhist belief in Japan, animals are included in the reincarnation system. Animals were believed to be born in such forms because their previous lives hadn't afford lives as human beings in their next lives. While faithful domestic animals and pets might reincarnate into human beings in their next lives, human beings who had problematic lives might reincarnate into animals in their next lives. Even after the war children were told that they might turn into some other animals in their next lives if they don't behave better, and old people would hope that their beloved pets might have chances to be re-born as human beings someday.

Generally speaking, it seems trivial to determine whether an object is a living creature or not. If it is so, it should be possible to answer to the following questions. What is the essence of the Greek "anima" which divides an organism from an inanimate object? What do we have as proof that some object is living, or not living?

According to modern biology there are no clear definitions of life. Each of those conditions which 'gives life' to an organism either also apply to lifeless objects, or do not apply to all living organisms. This only allows us a definition with a list of conditions that is not necessarily fulfilled completely.

Here lies one of the reasons why Artificial Life would exist: if there are existing organisms which do not fulfill ALL of the conditions for life, there is room for artificial systems which fulfill equal numbers of conditions to be thought of as a form of life. This allows a vast possibility for an artistic approach to life. Another reason is that unlike other fields in science such as mathematics, physics or chemistry, traditional biology only deals with historic development which traces the paths which life forms took on earth. It is a research area which discards entire worlds of possibilities (if there were not an ice age, if dinosaurs were not extinct, etc.) and bases itself on what possibilities brought the world as it is now. Then is it asking too much to bring into existence a meta-biological world which envelops as a local version, biology as framed within earth? As Euclidean geometry within non-Euclidean geometry, Newton's Mechanics within the theory of relativity, Artificial Life within Life. Such an argument is not out of order. The subject of Artificial Life is the "Life as it could be" as proclaimed by Christopher Langton in the First Artificial Life Symposium in Santa Fe.

Creating a Virtual Life

Within the computer itself, there are no mechanisms to discriminate between reality and imagination.

If a logically correct, yet unrealistic world is created (as examples, a different number is assigned to a parameter as opposed to a real number in the equations which simulate reality, or the equations themselves are transformed), it becomes the border between simulation and imagination. As the programming of a realistic 3D space computer graphic simulation is not much different from that of an imaginary space, such alterations which convert reality into non-reality are instantaneously possible with relatively simple changes in algorithms or parameters. The phenomena of these type of would-be worlds is well described in "Would-be Worlds" by John Casti[8].

The French zoologist and artist Louis Bec simulates an ecosystem of "theoretically possible" creatures. These imaginary creatures are designed using a model

---

The most painful part of this religious belief was that women were ranked under men (yet above beast, luckily) in this reincarnation hierarchy.

In short, traditionally there was no crucial boundary between lives of human beings and lives of other animals in Japan. It is often said that Japanese way of treating pets or robots is different from that of the West. The above explanation will give an explanation from the tradition.

[8] Casti J (1997) Would-be worlds, Wiley, New York

that fulfills necessary requirements to survive in the given environment, and are rendered using computer graphics. The colorful creatures swim in the virtual sea with beautiful movements which are correct in terms of fluid dynamics. Changes of parameters produce a variety of species which are different in shapes or patterns yet structurally similar, as we see with real creatures. Louis Bec carries the experiment not only for aesthetic purpose but with the aim of constructing "a more general zoological system which fills in the holes left by evolution"[9]. The amount of conditions of life rendered in his imaginary ecosystem is yet partial – mainly those that deal with the form and the motion which is the result of the form. However powerful they have become compared to those we had ten years ago, computers we have at hand are still far from good enough to deal with a life or an ecosystem in a comprehensive manner.

That is a reason why an artist working with the concept of life would focus on one aspect of life rather than trying to handle more than several of them, besides the fact that an artwork is usually more convincing when it contains a clear idea rather than splitting itself into different factors. Actually visual reality is one of those factors. Realizing a visually realistic image and behaviour of a life is by itself a great task in computer graphics if one thinks of the elements required in achieving such task. Form, color, texture, and realistic motion or deformation which should be based on the structure of the body and follow the dynamics, to name with.

Reality: Appearance or Algorithm

One popular misunderstanding on what might be called A-life art lies here. Sometimes viewers complain about the visual reality of the creatures in those pieces, comparing it to that they see on the wide screen of a Hollywood film. While the most important element of those computer generated animals or monsters in films is the visually convincing reality, the essence of dealing with life and bringing up a new point of view towards life in an artwork is not in the visual reality. In art the concept of life does not even have to take a life-like form. Life can be a more abstract entity. This is a point that art and A-Life share.

The way a biologist focuses in simulating life is also different from the approach of an artist. A simulation that a biologist carries out does not need to be visually realistic as well. But the essence of a simulation lies in its algorithm; the way it represents the reality. In other words, the result of a simulation does not necessarily have a realistic appearance, but the procedure should be logically correct. It is somewhat similar to the difference between genotype and phenotype. Rendering can give different appearance to a model.

"Polyworld" research project which was carried by Larry Yaeger after the concept of Alan Kay is an example of such case where visual representation of "life" – each of over three hundred creatures were represented by a simple polygon – was of secondary importance. The main focus of this ambitious research project was in simulating an algorithmically realistic autonomous ecosystem. But the

[9] IMAGINA 93 Proceedings

behaviour and the entire movement of the polygon-based creatures was interesting, like a choreographed dance of snowflakes.

In fact, as is often mentioned, the nature has its beauty. A successful simulation that conveys the way the nature crafts life could bring in an aesthetically interesting result. Apart from dealing with life, fractal geometry and its application in visualizing natural phenomena is a well known case [10]. On the other hand, rather than arbitrarily modeled or rendered shapes, or life-like looking objects which are modeled through mimicking and patch-working the outlook of existing creatures, forms and movements created upon the analysis of laws of nature would bring amazingly interesting or beautiful, convincing results for artists. Artists such as Yoichiro Kawaguchi and William Latham have proven the quality of logic within designing forms and motion. Kawaguchi's approach made a great influence on the development of computer graphics together with the contribution from a researcher/artist Alvy Ray Smith who started using particle system in computer graphics which is widely used now (Smith co-founded PIXAR company later), and aforementioned Mandelbrot and Voss, along with other researchers and artists who worked in analyzing nature's laws to bring it into art. Since then methodology in representation of natural phenomena, which used to be difficult by geometric modeling and would bring unrealistic outcome, switched into procedural modeling and later to more science-oriented physically-based modeling, dynamic simulation, and others. Art and science surely meet in this field.

When Christopher Langton first held the Artificial Life Symposium at the Santa Fe Institute Studies in the Sciences of Complexity in 1987, those gathered included researchers and artists of computer graphics who created imaginary life systems using computer graphics, or who experimented with simulation of birds in flocks and fish in schools for the sake of use in computer animation [11], besides researchers in life science including the aforementioned Richard Dawkins.

A-Life is a search for life using approaches which are not necessarily bound to reality. Here is the key to the necessity of the involvement of art in this new field of research. While Hieronymus Bosch created a 2D world in which strange creatures resided, the currently available interactive computer graphics and virtual reality allows us the possibility of being thrown into an autonomously existing ecology of his beasts. Within the computer, processes such as principles of life, physiological responses of humans, herding patterns of animals, ecology, and evolution of species expands beyond human predictions, and artists now have access to those results in real time using computers.

Automata

From the days of Pygmalion of the Greek mythology, who had the gods transform into life the marble statue which he himself carved and adored, the human

[10] The mathematician Benoit Mandelbrot and the computer graphics researcher Richard Voss are the major contributors in this field.

[11] Specific participants included Peter Oppenheimer who used simplified plant life cycle models to create imaginary plants, and Craig Reynolds who developed the models of bird and fish migration (his VOID algorithm has been used in films such as "Batman Returns" and "Cliff Hanger").

race has been infatuated with the creation of life. The first machinery in the form of human being that we can see the trace of is the mechanical serving girl by Philo of Byzantium[12]. The stories of fictional or non-existing creatures, life-forms in another space/time dimension are common themes across the board in mythology, science fiction, all of literature as well as art and film.

Today, such creation of life which used to belong to fantasy has been realized in blandly unromantic forms such as genetically engineered vegetables[13]. On the other hand, we see organism-like behaviours in robots with artificial intelligence, or computer viruses, those things which we do not normally consider as 'living'. In fact, the definitions arising from modern biology do not automatically exclude computer viruses or robots. In other words, there will always be exceptions to the definition of an organism such that validly existing life forms are excluded or obviously artificial forms fulfill the necessary condition. We have found that the definition of life is blurred by its nature.

Hoffman's story of Coppelia reflects the increased curiosity toward mechanical dolls (automata) which were extremely popular in Europe from the mid-eighteenth to early twentieth century[14]. The precision machinery of those days had the mastery to make people believe in such stories.

The automata developed during these years were not restricted to mechanical dolls but also included spontaneously moving devices. The mechanical dolls used combinations of gears like those used in cuckoo clocks and music boxes such that their movements are convincing facsimiles of a living human being (or monkey, bird). The duck made by the famous Jacques de Vaucanson in 1738 cleverly took advantage of the elasticity of the then new material, rubber. Its realistic simulation of the physiological workings of the bird ingesting food, digesting, and defecating was so beguiling that the bird was displayed on view for money. The most marveled process of turning the food into feces was revealed to be a trick. But this was only after more than one hundred years later by a famous magician Jean-Eugene Robert-Houdin who happened to repair the duck. Robert-Houdin had experience of watch-making as well as showing of automata (or apparent automata – a contrivance which appears to be an automaton though the actual mechanism was the workings of the magician). In 1845 Robert-Houdin was asked to

[12] Fascination with the mechanism can be seen strongly in Greek culture; the sketches of automatic mechanisms thought to have been designed by Hero of Alexandria, the mention of the working spaces of the gods of blacksmithing as well as the existence of the wooden horse of Troy in the Iliad .

[13] Actually these creations are chimeras of already existing species which are merely artificially manipulated with foreign DNA, thus, different from a living form created from a complete void.

[14] There were several celebrated automata makers in Japan as well. Japanese learned the technology from China in 7C. Introduction of mechanical watch from Europe made it possible for the manufacturers to learn more about the technology and apply it to different types of automata. Japanese automata developed mainly in and around Nagoya area. Sophisticated large-scale automata (automatic dolls) became the delight of festival cars in regions around Nagoya. Automata were made for in-house use and for attractions. "Tea-serving Doll" and "Shooting Doll" are such examples. One of the most famous automata makers was Hisashige Tanaka who later founded a company of precision machinery which developed into Toshiba.

Fig. 1. Vaucanson's automata: the Flute Payer, the Duck and the Flute and Drum Player

repair one of the wings of this duck, and revealed the trick in his book which he published in 1868. Due to this scandal Vaucanson's duck was discredited and was left to ruin. There only remain a drawing of its mechanical design and a photography of the wreck. However, Robert-Houdin himself was also using similar techniques[15].

The physiological simulation configured in Vaucanson's duck was more the exception. Amongst the numerous dancing, smoking, performing automata (initially the duck was also displayed with the "flute player" and the "drum player"), there were attempts in simulating intellectual activities such as the Harpsichord player, the Writer, and the Draughtsman constructed by Jacques-Doroz father and son, Pierre and Henri-Louis. These automata made a great success when they were brought to Paris on 1775 and then to London. Gustav Vichy's writing clown (circa 1900) is a later example of an automaton simulating an intellectual activity.

---

[15] "Magician and Film", "Automata et Bagatelles" pp.15–16.
The magician also opened the Theatre Robert-Houdin. During this time the magic of the magicians, hand in hand with the popularity of the automata, had become an important part of the theatrical stage set. It was the time that magicians started introducing scientific discoveries such as electro-magnetism. This parallels the transition from magic to optical illusions, and the eventual transition into film while magicians such as Georges Melies became involved in making films. They originally started using the medium to shoot their stage magic so that the film can be distributed to different theaters to meet the increasing demand of their shows, but eventually discovered the optical magic and founded the basis of special effects today.

Fig. 2. Von Kempelen's Chess Player showing its 'mechanism'

The most famous of all is the chess player, or the "Iron Moslem" (because of its outlook and the mechanism) crafted by von Kempelen in 1769. It is said that it beat Napoleon, but lost under a clever trick played by Catherine the Great of Russia. However, though the transmission of the precise replacement of the chess pieces was an ingeniously designed machinery, the thinking part, i.e. the one who determined which move to make was a little man hidden in the box[16]. Von Kempelen had also researched the mechanism of vocalization and designed a machine which he claimed to be able to correctly pronounce more than one hundred words. Goethe, who happened to see the vocalizing apparatus noted that it "was able to pronounce several childish words very nicely[17]. Kempelen was not the only one who took on the challenge of creating a voice box. Etienne-Gaspard Robertson who is known for the Fantasmagorie also invented a speech machine called "Phonorganon" in 1810 in the shape of a child[18].

By the turn of the century, these automata lost the status of theatrical showpieces and displays in wealthy homes. As a visual commodity, they were

---

[16] This doll was made in 1769 and was called by the name "The Iron Muslim" because of the attire of the doll sitting in front of the chess board.

[17] Kempelen W Ritter von (1791) Mechanismus der menschlichen Sprache, J.B. Dagen, cited from "Automata at Bagatelles", Vienna

[18] "Automata at Bagatelles"

passed down as electrically operated dolls in window displays, and as a technical contribution, they serves as precursors to automated vending machines and mass production of "talking dolls" by Edison. French writer Villiers de l'Ille-Adam gives Edison a major role in his novel 'L'Eve Future' which was written in 1886. In this book Edison creates a perfect mechanical automaton – a woman of perfection with intelligence – out of pure friendship for a young British noble man. But in reality it was Edison who was a man of ability not only in invention but also very much in business, who ruined sophisticated automata by promoting mass productive machineries and new medium of entertainment such as cinema.

Life as a Machinery

These automata and machineries were the direct results of our desire to mimic life. Different elements of life, from a purely physiological activity such as digestion to a mental activity such as playing chess, are represented in each of these automata.

What brought such fascination for automata, and why a fake such as the chess player could make people believe?

The automata had captured the people because of the arising belief of the times that all phenomena could be analyzed by logic and therefore can be simulated by technology. The development of machinery triggered by watches and then steam engines on one hand, and the advances in the sciences heralded by astronomy and medicine on the other hand, created such atmosphere of the time. The elucidation of the circulatory system by Sir William Harvey turned the divine mysteries of life into a mechanical system of flowing blood and became the basis of Descartes' Human as Machine.

It was the time when for the first time in our culture simulating life was regarded as a possible and reasonable idea. People could believe in the thought that different functions of life could be already realized in these automata, and thus integrating such functions to construct more sophisticated humanoid would become possible one day.

Descartes proposes a more objective standard for the definition of life. He thought of animals as one kind of automata, however it is interesting that he chose "that they use words and are not restricted to a patterned reply, instead can communicate in response to the situation", and "that they do not need an apparatus for each action and a single device suffices for all actions" as the ways in which to differentiate between a well-crafted automaton and a human being[19]. Later Alan Turing used the same line as Descartes' first condition for his 'Turing Test'. The second condition is in the same venue as the statement by the biologist Francois Jacob who compared engineering and bricolage to the nature of evolution. Since there were no such precise automata nor computers with such sophistication, the conditions of Descartes were in fact a part of an ideological model like in case of the Turing Machine.

---

[19] Descartes, "Introduction to methodology"

From Engineering to Bricolage

There have been major changes in the fundamental theories of evolution and ontogeny which resulted in the decline of the traditional determinism. Instead, ideas such as chaos, complexity, and emergence have become the key issue to life science. While it is the result of recent researches, it should be considered as a part of the inevitable flow of the history of science which can be interpreted as a series of discoveries that step by step pushed our species and planet aside from the privileged center of the universe. Meanwhile the change did not seem to be that drastic from Asian way of thinking where the traditional philosophy had a different approach to the status of human beings in relation to the rest of the world, as mentioned earlier.

Such change of our concept interestingly coincides with what has been taking place in our culture, even in the industry.

The theological statement that we must have been designed by God because it was impossible to imagine that such an exquisite mechanism could develop by itself, reflected the general feeling of the time when engineering was about to launch. The argument that William Paley made in his book "Natural Theology" in 1836 was convincing enough in the époque when sophisticated automata had been developed out of the watch-making technology. He "proved" that human beings must be the result of a design work, saying that when we see a stone on a road we recognize that it might have been there without any specific reason, but if we see a watch instead we immediately realize that someone should have dropped it, because we know that such a sophisticated mechanism does not grow naturally without any work and a clear intention. But now we know that it was a blind watchmaker who made us, not the God with a marvelous design skill[20].

There is an increasing questioning of the faith in engineering thinking which had pushed the world forward since the industrial revolution. As represented by expert systems, the top-down paradigm of artificial intelligence has come to a roadblock. Thus the attention is focused on the bottom-up paradigm which endows the computer the flexibility to adapt to the environment. There is a demand for such 'emergent' systems in the industry such as designing highly integrated circuit, as what engineers can think of has already reached the limit. It is a field where genetic algorithm has been applied for an extremely practical purpose. Such industry, which produces the core technology in computing, now seek for the breakthrough in emergent development, or, designing without blueprints.

French molecular biologist Francois Jacob relates the difference between the goal-oriented method of engineering and the emergent working style of bricolage to the nature's way of evolution in a speech entitled ' Le Bricolage de L'Evolution'.

In engineering one tries to accomplish the best result by designing for a specific goal, preparing a blueprint, and using brand new materials and tools one chooses for the specific purpose. It could be the most efficient way in achieving the already known goal. On the other hand it can realize only something that the human

[20] Dawkins R, "The Blind Watchmaker"

mind can foresee. Also, the blueprint should be modified each time when the environment changes.

Traditional concept of art making was not of Jacob's, but rather of Paley's. A masterpiece of art should be carefully thought of and designed by the artist, painted or sculptured, then installed on the wall or on the floor, and should be kept in its original form. A piece should be the genuine result of the artist's creativity and it should remain unchanged. There is an invisible (sometimes even visible at an exhibition) fence between the piece and the viewers. But is it the only way of making or appreciating art?

Gene, Meme, and Originality

Even if an artist thinks that her/his piece is the result of entirely original imagination, one always owes the source of imagination to one's experience, things one had seen, heard, read, etc. It might sound extreme if someone says because of this any work of an artist is based on a kind of shared memory or shared experience.

But in fact such point of view is possible and we are finding more works of artists based on the sharing of imagination or information. The possibility of global exchange and sharing of information on the network is about to change our culture. The key concept of network art projects by artists such as Akke Wagenaar or Muntadas is that the artist designs a system where people can offer and exchange information on certain themes.

Wagenaar's "Hiroshima Project" was conceived when the Smithsonian Institute had to cancel its exhibition proposal on atomic bombs. The project is an internet web page that guides visitors to any piece of information or opinions regarding the subject, from any point of view. Rather than producing junk information on the network an artist can use her/his intuition and the skill for visual communication in designing a system to develop our knowledge through exchanging information, Wagenaar says. "The File Room" by Muntadas is also an art project in the form of a database creation. Any information regarding censorship is collected and "filed" on the network, from the case of Plato in the "philosophy" drawer to Karl Lagerfeld in "fashion". The idea is to bring up pieces of information scattered worldwide onto an open database so that such information can be accessed, circulated and used by anyone. A piece of information or knowledge should not belong to just one person; it produces new information or knowledge when it is combined with other information.

Of course it is a basic concept of publishing or making archives. But it can produce much more dynamism when brought onto the network where its global consensus lies in sharing ideas and information. And this arising awareness toward the possibility and necessity of releasing information and sharing our intellectual resources such as knowledge and imagination has been discovering its connection to "gene" as a metaphor, or "meme" as proposed by Dawkins. On one hand, though not yet much known in the West, when we see works by Japanese artists we find that such concept nurtured under the influence of the recent researches in the life science are integrated into Asian (or rather Pan-pacific) traditional philosophy, finding a resemblance in it.

A Japanese network art project named RENGA which is conceived and carried out by artists Toshihiro Anzai and Rieko Nakamura shows how the nature of digital technology, especially the network, can be related to genetic metaphor in the context of Japanese culture[21].

RENGA originally means Japanese traditional linked verse which developed into HAIKU later. RENGA here is a word play. REN means link or linked, GA in its original terminology means song or poem. The same sound GA also means image(s), with a different Chinese character. RENGA is carried as follows. One artist will prepare an image. The image will be sent to the other artist via network. The other artist will modify the image freely, turning it into his/her own work. Then the image will be sent back to the other. The session continues until they feel the series is saturated.

While the process itself can be seen as a collaborative painting, the concept of RENGA is deeply connected to the nature of digital technology and the idea of originality. The transition of the idea of originality with digital technology is a well known fact (and might be a concern from the business point of view), but it can be viewed from a different point of view in our Asian culture. It will be discussed later.

RENGA is a process to share someone else's imagination. Instead of limiting the source of imagination to that of oneself, there are different elements that grew out of different imagination. On the other hand, one is obliged to start from what is there, and produce something different from it according to one's own imagination or necessity of expression. This is a bricolage, not an engineering. A RENGA artist (except for the first image to be used as the "seed") does not start from drawing a blueprint of what she/he plans to produce. It is interesting to know what it brought to a participant to the International RENGA session that took place during SIGGRAPH 94 in Orlando. "I usually don't paint such piece. But the image sent to me provoked something in me that I hadn't realized until then. I never thought that I had such imagination. I couldn't help painting this piece. It was me who painted – but it was not just me. " As in case of bricolage or in evolution, referring Jacob's statement, an artist might discover an image that he was

---

[21] Anzai and Nakamura have been using computers for painting and networking professionally since 80s. In early 90s there were a few networks including Anzai's own that functioned as meeting places for media artists and researchers where discussions on the concept of originality in digital era continued, along with MOO and MUD which was carried without even knowing how such ideas were called.
Many of the artists who joined Digital Image (the largest artists group using digital medium which was founded in 1990 and had its first exhibition on 1991, the co-founders includes the author herself) overlapped with the members of such networks which were inter-related. Because of this network art projects among these artists launched on 1991.
A three day international RENGA session was organized connecting the EDGE at SIGGRAPH 94 and artists on network in Japan. An image would go through the modification at SIGGRAPH with artists visiting the space, then the final piece of the day will be sent to Japan at the end of the afternoon. In the morning in Orlando the image sent back from Japan will be uploaded and the same procedure is repeated. Another international session was held during ISEA 95 between Montreal and Tokyo.

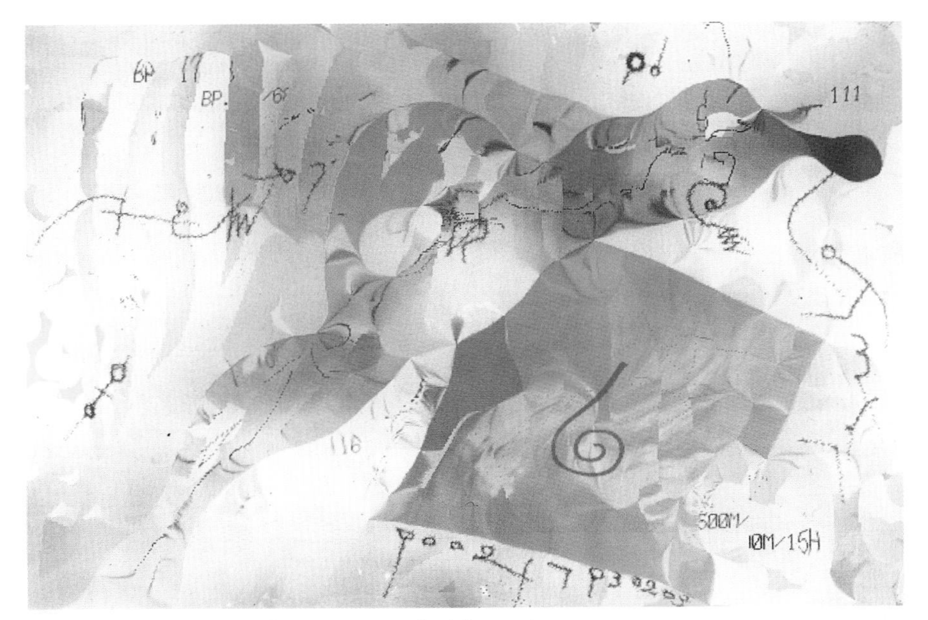

Fig. 3. RENGA. © Toshihiro Anzai and Rieko Nakamura

unable to foresee. In fact it is interesting to see RENGA sessions. Colors and motifs originating from different imagination would survive, transform, cross or melt together while weaker elements might disappear. The images produced through each session are different from what each artist would make by himself. Anzai and Nakamura developed a concept of a "gene pool" of imagination. In their "Ninohashi RENGA" (1994) project supported by InterCommunication Center and "Visual Jazz" (1995) which was performed realtime such "gene pool" was prepared and used.

Gardener of Eden

The concept of gene is thus treated as a metaphor in case of RENGA from artistic point of view. It has nothing to do with genetic algorithm. British artist William Latham started using computers originally to visualize his hypothesis that any existing form should be explained as a combination of primitive forms and deformation that follows. Latham had made hand drawings on wall-size sheets of paper to show the map of the evolution of form [22]. He started working with researchers at the IBM Science Research Center and found that by applying the hypothesis proposed by the British biologist Thompson D'Arcy (which was also a basis of Yoichiro Kawaguchi's approach) to computer generation of images a simple algorithm can produce a variety of organic shapes by selecting different parameters in modeling and rendering. By interpolating between different forms which were composed of the same number of components, Latham made amazing series of

[22] William Latham "The Evolution of Form" O Museum

animations where organic forms change from one to another. Such animation, such change of form is impossible to make without using computer and applying algorithmic approach, because it is beyond the capability of our imagination, or capability of designing. Latham says that this is the reason why he uses computer. He puts his position as a gardener of god[23]. It is Latham, the artist, who makes the choice according to his aesthetics, but the creation is not a hundred per cent on his own. He is a gardener who selects the right species and help them grow. And the selection is according to his aesthetics because this is an art project that employs computer technology for the creation of beauty, Latham says. His most recent project is a CD-ROM version of his program. A user can make a selection for a series of parameters to design a virtual creature of Latham. Or, is it a creature of the artist, or the user? The question is still open.

Beyond the Imagination

The power of genetic algorithm was most visually proven by Karl Sims when he produced the series of works including "Primordial Dance", not to mention his most recent projects dealing with the evolution of movements of virtual creatures.

It is a common place knowledge that any image on computer screen can be described by equations and numbers, but no one had thought that the evolutionary change of equations could result in eyes-taking development of images. Starting with simple equations, some of the images developed into organic figures which are difficult to imagine how they could have been calculated. As Sims is both a computer graphics researcher/software engineer and an artist, the process of evolution he made – or the way he helped his program to evolve – according to his own aesthetics brought about amazing results.

"Genetic Images" is an interactive version of the same system. That is, instead of the artist himself visitors to the exhibition would make the natural selection from the images by choosing several favorite images that will survive as parents to the next generation. When asked why he made the evolution process open to the public, Sims answered that he was curious to see other possibilities of evolution through other series of selection. Although his aesthetics – or the capability as the gardener – had been proved by the resulting images, he came to notice that the process of evolution was limited to the direction his aesthetics allows each time he made the experiment. The result of "going to public" was interesting enough when it was first shown in Paris, at Centre Pompidou. Some images which were of different types from what Sims had achieved by himself were discovered. An evolution of images would grow a different branch if it takes place in a different environment, i.e. with a different user.

From the point of view of art making, this work illustrates the interesting possibility that interactive art offers to the artists not only to the visitors. An artist can design a system instead of designing the final product, to offer a joy for visitors to participate in art making, but at the same time also to merit from one's own work

[23] Ibid.

to see what the artist him/herself would not have imagined, with the assistance from the visitors. It is not an easy task to design an interactive piece with an efficiently working system and good user interface to bring people in, but when it works artists can also enjoy the results.

Christa Sommerer's and Laurent Mignonneau's recent work "GENMA – Genetic Manipulator" is a 3-dimensional design engine that allow users to make selections from parameters to create a virtual insect-like animal. The fact that it is real time, which means that the 3-dimensional configuration of a creature takes place in front of the user's eyes in fully rendered image, carries a startling experience. It is not necessary to discuss here about their earlier pieces – I have already written very often about the way the artists integrate the goal and the design of their pieces, employing natural human interface, and they write about their own works. But such possibility of the real-time gardening – or what should it be called in case of animals? – opens up new field of imagination.

Artists are the specialist in visualizing imagination which go further than most of the ordinary people. But an artist might have a dream to create what is even beyond his/her own imagination. Automatism was used by DADA artists, it tried to capture what lies below the consciousness. Happenings and the use of media technology were methods taken by Fluxus members. There are different approaches to achieve the goal, typically from psychological approach and technical approach. Interactive art is forming an arena where these different methods meet and are integrated. It is natural, if we think of the fact that they are originally different activities that belong to the same human species.

Genes in Japanese Poetry

Going back to the issue on the concept of originality, Japanese short poem or WAKA contained gene-like concepts quite early. RENGA developed from WAKA. Actually there was much influence from Chinese literature in the foundation of such poem, but it spread among normal people[24]. First official anthology of Waka includes many poems made by normal anonymous people from all over the country. These short poems were not necessarily the expression of their feelings but were also used as a tool of communication either to convey polite greetings or passionate affection, or even political messages.

But how can a poem with only 5-7-5-7-7 syllables (which might be a composition of six to ten words) convey so much meanings?

Actually it was exactly the reason why Waka served as a useful tool of communication. Instead of being a poem that stands by itself with meanings that come directly from its components, a Waka would consist of words that have different meanings behind them. In a sense, Waka was a multi-textural non-linear form of literature. Because of that a Waka could mean much more than what was literally said.

What would happen to the idea of copyright or originality in a society where poem is a part of communication and is recycled as a part of a database of meta-

[24] Takaaki Yoshimoto "Tanka-ron"

phor? A good poem will be cited, not as a whole but just as a part of it. Citing the whole poem does not make the match of intelligence. A word from a good poem would start to have its own meaning, and would be used over and over, gradually changing or enriching its original meanings. In a sense, it becomes a gene in the literature. A good, useful gene will be used, modified, and keeps on living in different forms. It could become a makura-kotoba, for example. Finally the origin of such gene might be forgotten because it becomes a part of the environment. On the other hand, a bad or banal poem would only remain as it was, without being referred or cited. A poet should feel happy if a part of his/her poem is used in a different poem by someone else. An appreciated poem will be recorded and printed with the name of the poet on the official publication from the court (such publication was regularly accomplished) besides being remembered by others. But at the same time it should be decomposed and re-used by others to prove it's metaphorical strength. It is difficult to say either such tradition influenced and prepared the way Japanese attitude toward originality, or it comes more deeply from Japanese or Asian culture. Judging from the fact that Asian countries share similar attitude toward copyright, it is likely that either Asian culture or Pan-Pacific culture cultivated such tendency.

Waka gave birth to another important form of Japanese poem RENGA which developed into HAIKU later.

Self and Other

Video artist Takahiko Iimura made an installation in the early 70s with two video cameras and monitors shooting and showing each other, asking "who am I " and "who are you" to each other[25]. It is a most basic question in philosophy as well as in art put exactly in its right context using media technology. It is an example of what media technology has brought to the artists – the possibility of extracting such question (life's big question...) and show it in the most explicit way. Interestingly enough, in this case, the question is one of the major issues in modern biology as well.

Japanese traditional notion of self and other is different from that of the West. The difference reflects in media art as well.

Noriyuki Tanaka is known for his collaborative works with Shinsuke Shimojo of the University of Tokyo which brought art and cognitive science together[26]. Creating a space where a visitor would realize his/her unconscious ego, expectations, or automatic way of cognition, is the purpose of Tanaka's artwork. With Tanaka's two images uploaded on the network users of the net created different resulting images. One of the most interesting works came from an architect who by using 3D computer graphics made a virtual installation of Tanaka's works. For Tanaka the project was interesting to experience how his imagination would go through changes through other people's imagination. The resulting images are mixtures of imaginations of the artist and the users. The artist experimented the

[25] Takahiko Iimura Newsletter

[26] Noriyuki Tanaka "PAGES"

idea even further in his CD-ROM publication then. "The Art of Clear Light"[27], a CD-ROM which contains photographs Tanaka took in different places is not just a photo album of an artist. It contains a piece of software that shuffles the images in the CD-ROM and overlay each of them on top of images from a different folder on the computer. A user is invited to bring in his/her own photographs or drawings into the folder so that each image will be mixed with the artist piece of work.

Art projects to bring an experience to step beyond one's self are carried by artist Kazuhiko Hachiya as well. In his work "InterDiscommunication Machine" Hachiya made a parody of InterCommunication Center(NTT/ICC) in its naming, and a parody of high-end expensive virtual reality in its system designing. ICC is named after the concept that the new media technology would connect people creating new channel of communication. InterDiscommunication Machine shows that technology can serve to cause discommunication among people. Actually the aim of the work is to promote communication between two visitors by physically relocating the normal communication channel. With this piece of art, each of two visitors(users) is asked to wear a special equipment. It exactly looks like a kind of

Fig. 4. InterDiscommunication Machine. © Kazuhiko Hachiya

HMD with a screen and a set of headphones with a transmitter on one's back. What you see is only the screen. What one sees on the screen is the space in front of him/her from a different point of view. It is, in fact, the space seen from the OTHER person! What you see is what the other person sees, what you hear is

[27] Noriyuki Tanaka "The Art of Clear Light" DIGITALOGUE

what the other person hears. A small video camera and a microphone on top of the helmet shoots the supposed view and collects the sound around a visitor. The image and sound are transmitted wireless to the other person's screen and the speaker. Though the technology used is very simple, this system allows the exchange of one's view and the soundscape with that of the other person. In a sense it is an extremely low-tech virtual reality. It is difficult to imagine what would be such an experience unless you try it. You should look around for yourself on the screen to reach the other. If you see yourself it means that the other person now sees you. You should understand the space around you by guessing what the other person is doing – because you can see it only through the other person's eyes. Shaking hands is already a big deal. People say "try to see things from other people's point of view". When it comes true, seeing things from another person's point of view is not that easy. Gradually you get used to a sense of having one's tele-existence that belongs to another person's coordinate system. It is a strange feeling to merge one's world into someone else's cognitive space[28].

Toward Cultural Bio-Diversity

It is not easy to define the relationship between art and science, or art and technology. What lies underneath of them is culture, but art, science and technology also play important roles in culture. Even with the research of complexity it would take time to make a working model of it.

Working and doing research in the crossing point of art, science and technology, having lived in different countries, I am more and more aware of the necessity of understanding the current status of these fields in relation to the cultural background to have a better use of different elements or resources we have, i.e. genes or memes we have altogether on the earth.

Japanese are well known for their efficiency in working in groups. It is why Japanese industry succeeded. On the other hand Japanese are not very good in making decisions on one's own responsibility. Seeking more value on collaboration rather than individual goal makes a totalitarian attitude easier, which was unfortunately proven in our recent history. Our tradition in putting less importance on personal right on art pieces should be a reason of the problem on copyright issue, such as copying software. But the same tradition might bring new possibilities on network, allowing free transaction of imagination. It is a different approach, coming from different background.

There is a word "bio-diversity". An ecosystem with a rich variety of species is more stable and would survive through drastic environmental changes. Species which had developed to fit the environment would have difficulties when something changes. Minor species might find the new environment more comfortable and would become prosperous. If there are no such species all species might die. As is known in case of the regeneration of forest, the ecosystem itself is the combination of different species.

I believe in the necessity of the cultural bio-diversity. Each culture has its

[28] "Ars Electronica Festival 96 MEMESIS"

own tradition. Through its history a society would generate its own culture; the whole complex of art, society, way of thinking, way of working, etc. which are connected together with the same backbone and nerve system. Every now and then there would be a prevailing culture that orients the global fauna, but when a big change happens to the environment it is such diversity of culture that helps finding the way to modify the rule and keep the global society adapt the new condition. And the environment is in fact changing rapidly and globally. The new way of communicating such as network is changing our way of thinking and way of living. We need to keep our cultural biodiversity so that we keep our capability in confronting possible confusion in our society. We need imagination to foresee our future. Art and science can work together to bring such imagination.

4. Artists as Researchers

Art ("and" or "versus") Technology – Some Personal Observations

Michael Naimark

Introduction – The Dualism of Art and Technology

"Art and Technology," like art-and-anything, addresses a dual agenda. To describe oneself as a conceptual artist, a feminist artist, or a video artist is to acknowledge a dualism between one's genre, politics, or medium and one's art. And like all dualisms, sometimes there is symbiosis and sometimes there is strife.

I believe in the existance of "pure art," art without any other agenda but the art itself. I became convinced when I met He Gong, a young Chinese artist, several years ago while we were both in residence at the Banff Centre for the Arts. He had spent the first part of his studies learning political art (by and for the Chinese government) and the latter part as an activist artist working against his government. He spoke English and had a strong background in contemporary art (from his dissident university instructors), but this was his first time outside of China. He Gong spent weeks in his studio working vigorously on a personal installation made of wood, rice paper, ink brush, and eventually, fire. He once said to me "I am so grateful. This is the first time in my life a can do 'pure art.'"

For some of us, sometimes, the path chosen requires channeling the passion and necessity of artmaking into particular issues, in my case – technology. It is my observation and belief that technology, particularly computer and media technology, has and will continue to have an increasingly profound effect on everyone on the planet. And that if artists don't jump in and pro-actively help shape these powerful new tools, it will be left by default to advertisers, the military, organized religion, and sex peddlers. Some of us believe the stakes are high.

That's been my attitude for the past twenty years, and I've had the good fortune during that time of working inside a variety of institutions with similar beliefs (or which at least tolerated mine). These places supported my own work and for this I am grateful. In fact, my projects could not have been realized without their help.

But it wasn't always a cakewalk. Sometimes it felt like the "art" and the "technology" forces were in opposition. This paper offers observations and reflections on some of these issues. The purpose here is to learn from the past.

MIT (1976–1980) – Is the Demo the Beginning or the End?

MIT Media Lab was a lively place for art and technology during the late 1970s, when I was there first as a graduate student and later as a Fellow at the art center, the Center for Advanced Visual Studies. CAVS focussed on environmental art under the direction of Otto Piene and its founder Gyorgy Kepes. The Film/Video Department, run by Ricky Leacock, was participating in all sorts of video experiments. Meanwhile, Nicholas Negroponte headed the Architecture Machine Group, which was well-funded and increasingly getting involved in media.

In 1977 I had this crazy idea to move a movie projector to mimic the original camera movement. I asked Nicholas for funding. He agreed, and I made a simple study by filming with a super-8 film camera on a slowly rotating turntable, then replacing the camera with a small loop projector. The result, which we called "moving movies," retained the film's original directionality and appeared as natural as viewing a dark space with a flashlight.

After showing this to Nicholas I said "great, now I'm ready to begin" and he said "great, now you're done." I was interested in exploring imagery and he was interested in the technical process. To confuse matters further, at that moment, we were just beginning a new project using one of the very first prototype laser-disc players. The idea was to film along pre-determined routes with stop-frame cameras and make an interactive system which allowed end-users some control over speed and direction. The project, called the Aspen Moviemap, wasn't intended to be an art project but dealt with some classic issues of visual representation. We all knew we were breaking new ground. I continued working on this project for the next two years, and since then made several other moviemaps. But I also kept working on moving movies.

For the next two years I built various camera and projector contraptions to move the image with better control, but then I felt like I had to decide: was I interested in building a new projector or in making an art statement? I opted for the latter, and over the next four years produced a series of installations reverting back to a simple turntable, but where I could concentrate more on the imagery itself.[1]

The Architecture Machine grew into the Media Lab and prospered, while CAVS increasingly struggled through the 1980s. I believe this split between the well-funded technologists and the struggling artists was microcosmic of what was happening in the US during this period. But more on that later. The lesson at the time was that demonstrating a novel idea was different than using it toward artistic ends.

Atari Research (1982–1984) – Everyone is Not Like Us

In 1982 the Atari Corporation, which was making an incredible amount of money on video games, decided to start a long-term research lab to look ten years ahead into the future of computing. They hired Alan Kay as Atari's Chief Scientist, who immediately went about rounding up a hundred mostly young people he thought would be "visionaries" for this task. Many of these young people were from the emerging MIT media scene, as well as a diverse group of others. Having already moved to San Francisco in 1980, I was brought in as well.

One problem I noticed is when you put a bunch of very bright people together to speculate about the future, they do just that: speculate. This can be dangerous, because it's easy to cut off the rest of the world and assume everyone is just like you.

After a year, one researcher, Bob Stein (who later co-founded Voyager, the interactive publishing company), did something noteworthy. He hired a local twelve year old boy to keep with him at all times a small portable tape recorder, and to record every question that came to mind over the course of several days. Bob's idea was to see what kind of questions everyday people might have, since we rarely remember most of them. This seemed like an important start if we were trying to understand how people in the future might use portable computers.

Bob chose a Palo Alto boy whose parents (both of them) were Stanford faculty. Virtually all of the questions he recorded were the sort whose answers could be found in an encyclopedia, straightforward educational questions. I wanted to respond in a way to both compliment and challenge Bob's work.

Bob distributed his transcripts to the Atari Research community. That evening I met with several anthropologist friends who after dinner, wine, and looking at dozens of maps and atlases, had converged on a plan. I would go to the remote northern mountains of the Philippines to visit a tribal culture called the Ifugao, a culture very different from ours, but where some people speak English. They are known for their ancient and spectacular rice terraces, for having been head-hunters, and for their strong belief in dreams. Two of my anthropologist friends had been there a few years prior, and wrote me a letter of introduction to someone they had met, a sixty-six year old Ifugao Shaman named Dionicio Immatong. I left the following day, and took with me a small portable tape recorder.

Several days later I found myself inside Dionicio's hut, where he read the letter by candle light and took me into his family and his home as a son. We set out to find a child to ask him to record what was on his mind as if he was interacting with a machine, just like the Palo Alto boy. We found a twelve year old Ifugao boy from the village of Paypayan named Patrick Tundagui. Patrick recorded every question that came to mind over the course of several days.

Patrick's questions differed significantly from the Palo Alto boy's. For one thing, Patrick made multiple use of the word "you," sometimes referring to anyone and sometimes to a particular person. He often questioned the certainty of hard facts, asking questions like "how do you know this is the smallest bird?" And he sometimes asked questions which were personal rather than encyclopedic, like "what is your problem now?" [2]

I'll admit this was a bit of a stunt on my part. And it was only a sample of one, so it's important not to read to deeply into any conclusions. But it did have the effect of shaking things up a bit back at Atari and reminding ourselves that not everyone is just like us.

Apple Multimedia Lab (1988–1990) – Educators and Artists are Different

Atari crashed in a big and ugly way in 1984. Many of the people resurfaced several years later at Apple and Lucasfilm. In 1987, a conspiracy of sorts was made between some of these people to convince both companies to start a multi-media laboratory. Neither company was willing at that time to commit to multimedia, but together they approved of the formation of the Apple Multimedia Lab, located in San Francisco, mid-way between Apple in Silicon Valley and Lucasfilm in Marin County. These were close colleagues of mine, and I was invited to help.

Our flagship project was called the "Visual Almanac," Apple's first interactive laserdisc made primarily for schools. I directed production of the laserdisc, which consisted of thousands of short sequences of still images, video clips, and weird stuff.[3]

I remember sitting in a meeting with several consulting teachers and listening to how they all tried to communicate so clearly. I became depressed: they were trying to communicate their ideas by saying everything in such an obvious and explicit way. This is not the way artists I knew operate; we seem to be more concerned with creating a feeling, an impression, a metaphor.

This distinction came to a head on a little piece I was producing for the disc, of a main street in Silicon Valley filmed by the State of California Transporation Department very much like the moviemaps I'd been making. They filmed one frame every 52.8 feet, or one hundred frames per mile, from a camera car. And they'd been doing it since the early 1970s. I selected an interesting hundred frames and made a split-screen version of their earliest film and their latest film, a "then and now" comparison of how things have changed over this one-mile strip in Silicon Valley.

Several colleagues on the project wanted me to add educational information about each of the buildings, but I refused, wishing instead for the visual impact

Fig. 1. Be Now Here. © (1995) Michael Naimark at Interval Research, USA

Fig. 2. Be Now Here – filming. © (1995) Michael Naimark at Interval Research, USA

of the material to stand on its own. Then they said "you can add it, and since it's interactive the user doesn't have to see it" and I still said NO. I felt this was a trap of sorts.

At any rate, I was left with the impression that educators and artists have different intentions. Maybe "intentions" is too strong a word here, since both educators and artists might say their intention is enlightenment. But even so, educators tend to spell things out in a more literal way while artists have less of a problem with ambiguity.

San Francisco Art Institute (1989–1990) – It's a Small Art World

I may have been particularly sensitive to this distinction since I was also teaching at the San Francisco Art Institute, a landmark institution for contemporary art, a cutting edge place. I was teaching a class called "Virtual Evironments" and asked the Apple lab if we could borrow a Macintosh, a laserdisc player, and one of the then-new little liquid crystal display video projectors.

The students produced an amibitious virtual environment of a restaurant we named "EAT," involving students performing as waiters and images of food (among other things) projected onto the diner's plate from a video projector hidden under the table. EAT was exhibited at various art venues, but it also showed at SIGGRAPH.[4]

The next year, my students produced a videotape parody of virtual reality called "Virtuality, Inc." It received a "Futures Scenarios" award at SIGCHI, the major computer-human-interaction conference.[5]

I realized that I was pushing these projects in the direction of the research community more than the art community, like making little "art bombs" and lobbing them over the fence into foreign territory. I must say I was proud of that. It was also great fun. I very much wanted the art to impact the research community.

But the fact is, almost no one at the Art Institute had any knowledge of these venues and saw little relevance. It was outside the art world.

Things have changed a bit since then. As the Internet, multimedia, virtual reality, and the Web have become trendy to the mainstream culture, they have become fashionable in the arts community as well. Nevertheless, making art for communities outside the art community felt like an uphill climb.

Banff Centre for the Arts (1991–1993) – Local Support for Global Activities is Vulnerable
The Banff Centre for the Arts had the most remarkable program for art and technology I'd ever seen. For one thing, it's in the Canadian Rocky Mountains, a most

Fig. 3. See Banff! © (1994) Michael Naimark at Interval Research, USA

Fig. 4. See Banff – filming. © (1993) Michael Naimark at Interval Research, USA

beautiful place. It's a large complex, complete with swimming pool, health club, and bar as well as food and lodging facilities. But most important, it's an art center, with hundreds of music, performing, and visual artists from all over the world together in residence. Inside this art center was a tech-based lab called the Art and Virtual Environments program which opened in 1991. It truly was a unique program.

The following year I began a project there called "Field Recording Studies," to explore turning real-world imagery into 3D computer models. That summer I was committed to exhibiting at Siggraph, and I literally flew into Banff 10 days before I was to fly out to Chicago with something to show. I had a simple idea and they had a superb team of technologist helpers as well as state-of-the-art SGI computers and video post-production facilities.

We produced a 360 degree panoramic computer model of a nearby landscape at dawn. The concept was to make one contiguous panorama by "tiling" together slightly overlapping still images. I collected these images with a small consumer video camera on a tripod, using traditional surveying tools like a compass and level. Back at the lab, the Banff Centre technical staff had prepared software to allow hand-positioning of the forty-two images we digitized off the videotape. The result was a three-dimensional model of panoramic dome.[6]

Everything worked out well. The Banff Centre was, in my opinion, the perfect place for art and technology work, particularly because of its international multicultural diversity. So what was the issue? The program no longer exists. A couple years later, the conservative party won in the Banff region, and funding cutting-edge non-traditional arts programs was among the first to go. The local conservative community didn't appreciate or find relevant the fact that the Banff Centre was known and active on a global scale.

Of course the Banff Centre still exists, and in many ways is still thriving, but the Virtual Environments program is gone. I'm hopeful this is a temporary situation. For one thing, the explosive growth of the Web is making all activities potentially global, and institutions will have to adapt if they have strong parochial views.

Interval Research (1992–) – Can Enterprising Technologists Deal with Independent Artists?

Funding for the arts, like most social spending in the US, had been very heavily cut back by twelve years of Reagan and Bush conservatism. By 1992, the US arts community was underfunded, heavily politicized, and to some extent, angry. During this same twelve year period (and for some of the same reasons) much of the high-tech community prospered. The cultural gap between high-tech entrepreneurs and independent artists had grown large.

In 1992 I was offered a research appointment at Interval Research Corporation, a new independent research lab wholly owned by Microsoft co-founder and billionaire Paul Allen. It's charter was to look five to ten years ahead into the future of computing and media, in a most general way. Unlike other tech labs I'd seen, this one seemed to really believe in having artists and other diverse elements as members of the research staff.

I'd been completing the last phase of my Banff Centre project on field recording, and Interval's head David Liddle assured me that art will be an integral component in this new lab. I could continue to work as I was and make something exhibitable. The result was called "See Banff!," a stereoscopic moviemap (the first ever) about landscape, tourism, and growth in the Canadian Rocky Mountains. It was filmed with twin 16mm cameras and displayed as a single-user experience housed in a cabinet resembling a century-old kinetoscope, with a crank on the side for "moving through" the material.[7]

One particularly fruitful collaboration that came out of the Interval community was with the computer vision researchers. I learned they were also interested in basic elements of visual perception, perspective, and presence, and together we nurtured a symbiosis. The footage I produced for See Banff was also made with them in mind. They were amused, I think, to have an artist-type supplying them with material which they felt was unique and valuable. The fact that it was not simply "views of the parking lot" was gravy.

Over a two year period, we all did pretty well. Working with my Interval colleagues, we designed an experimental camera system. I had several weeks of filming as I like best, open-ended and with participation by local community people, and made an installation. My computer vision colleagues got some unique footage and made some striking new imagery. It turns out we also got a patent out of it, something totally unanticipated when we began.

So beginning the next year, in 1994, I proposed we try it again, this time working with representing "looking around" the way the Banff project represented "moving around." We put together another experimental camera rig, this time using two 35mm motion picture cameras for stereoscopic 3D, running at sixty frames per second for unrivalled fidelity. Like my earliest work, the cameras would rotate on a motorized tripod to capture the entire panorama. And I'd work with local community people, but this time in collaboration with the UNESCO World Heritage Centre based in Paris. With their endorsement, I'd take the camera system around the world to film in endangered places. Finally, the footage would be shown with the viewers standing on a slowly rotating floor, which rotates in sync with the imagery. The effect is illusionistic, like the feeling when the train next to yours pulls out of the station and you think your train is moving. The final installation is called "Be Now Here" and was produced for the Center for the Arts Yerba Buena Gardens in San Francisco.[8]

Again, I had managed to produce an art installation. And again, my colleagues got unique footage for their research. And it turns out again, we also got another unintended patent application out of it.

It also turns out that we had inadvertantly helped another cause. One of the endangered places I filmed was Dubrovnik, the medieval Croatian town near the Bosnian border. It had been heavily bombed and was still in a state of war. Dubrovnik had a newly created Web site, and its designer Enver Sehovic, a professor and former President of the University of Zagreb, had a vision: to save Dubrovnik through a Web presence. Professor Sehovic helped me get in and out of Croatia during the fighting with my five hundred pounds of film gear.[9] Shortly after the installation opened in San Francisco, Sehovic emailed me that my project was "extremely important for Dubrovnik" and that he was coming to see it to help convince the Croatian government to support his project, to show that he's not just a "dreaming professor." Perhaps not the primary concern for Interval Research, but things like this may happen when projects involve real world people and places.

So what is the problem? History. After more than a decade of technology entrepreneurs profitting while the arts community has been almost strangled, new bridges need to be built.

And perhaps the timing can't be better. The tech world is realizing that consumers don't buy technology for its own sake but for the experiences they afford. The word "content" has only come in vogue recently (and indeed, has entered the vernacular of the Media Lab and its sponsors). The toaster-makers have finally realized that people don't want toasters, they want toast.

So, can enterprising technologists deal with independent artists? I don't know for sure. There are potential problems, including issues of tolerance and compromise, of intellectual property and secrecy, and of artists being true to their hearts about motivations. Some of us believe the stakes are high.

References

1 Naimark M (1980) Moving Movie, Aspen Center for the Arts; (1980) Movie Room,

Center for Advanced Visual Studies, M.I.T.; (1984) Displacements, San Francisco Museum of Modern Art
2 Naimark M (1989) The Question Machine. Whole Earth Review 65: 54–55
3 Apple Multimedia Lab (1989) The Visual Almanac. Apple Computer, San Francisco
4 Naimark M (1991) EAT – A Virtual Dining Environment. Tomorrow's Realities Catalog, ACM Siggraph, Las Vegas
5 Naimark M (1992) Virtuality, Inc. SigChi Catalog, ACM SigChi, Monterey
6 Naimark M (1996) Field Recording Studies. In: Moser MA (ed) Immersed in Technology: Art and Virtual Environments. MIT, Cambridge, pp 299–302
7 Naimark M (1997) A 3D Moviemap and a 3D Panorama. SPIE Proceedings 3012
8 Ibid
9 See: http://www.interval.com/projects/be_now_here/trip.html

Images of the Body in the House of Illusion

Monika Fleischmann and Wolfgang Strauss

The world is an illusion -
and art a representation of the world of illusion.
Paul Virilio

Making the Invisible Visible

"What is important is what you can't see", stated Saint-Exupéry's Little Prince[1]. "Making the invisible visible" in order to research what is important. This is how we describe virtual reality in 1991. Moreover: "Compacting the diversity of the world into a single unit – the binding power of the idea and the fundamental picture"[2]. ART + COM conducts "research and design in extended dimensions" and sees itself as an "enterprise that endeavours to create a 'total art, science and technology work' which is devoid of boundaries and based on a common "digital code". Designers, artists, computer experts, scientists and specialists from the humanities are looking to pool their resources in order to comprehend the digital world effectively.

The speed at which information is conveyed and the quantities involved can be overwhelming for the individual. Can the arts work with science and technology to counteract this feeling of disorientation? We begin to examine light, acoustics, time, space, movement and perception through the capabilities of the computer, investigating these from the aspect of architecture, art, physics, computer graphics, video and design. The concept of 'invisibility' is one that is shared by each of the disciplines concerned. The financial basis for founding ART + COM (1988) as an independent research institute under the leadership of Edouard Bannwart has been created by the three large-scale projects "New media in urban planning"[3], "Design of comfort factors in (digital) space: investigations into light, air and acoustic space"[4] and "Visualisation of digital medical data"[5] which have

[1] Cf. Saint-Exupéry A (1992) Der Kleine Prinz (The Little Prince), Rauch, Düsseldorf

[2] Cf. the first ART + COM presentation brochure from 1991 designed by Monika Fleischmann: "Forschung und Gestaltung in erweiterten Dimensionen" (Research and design in extended dimensions)

[3] The project "Neue Medien im Städtebau" (New media in urban planning) is put forward in 1988 by Prof. Edouard Bannwart (architect) from the Berlin Senate for Science and Research and is funded by DT Berkom (subsidiary of Deutsche Telekom).

[4] The project "Gestaltung von Behaglichkeitsfaktoren im Raum: Licht-, Luft- und Schallraumuntersuchungen" (Design of comfort factors in space: investigations into light, air and acoustic space)

been funded by Deutsche Telekom / Berkom and the Senate for Science and Research.

On the Track of Our Lost Senses
New virtual communcation space need new techniques of illusion. When we reduce our multisensory abilities on a certain sense the virtual communication becomes another dimension. The telephone is so much accepted because the connected people concentrate on each other by only one sense – the sense of hearing.

Virtual Communication Space and Immersive Illusionary Techniques in which the body is confronted with new experience are main aereas of our interest. We overlay real and virtual worlds to create acoustic spaces, talking walls, echoes and acoustic distortions. We endeavour to find digital correspondences for light and air images such as reflections, movement, light diffraction and the dynamic incidence of light in its changing forms. Our study of reality and virtual reality is concentrated around the senses and their deception. While the idea of constant metamorphosis, symbolised by Hermes[6], and the concept of identity – which is also an important theme in the myth of Narcissus[7] – are other key areas of interest. It is our goal to bring persons "into contact" with the world, with each other and with themselves. We are therefore on the track of man's lost senses in a bid to restore these with the aid of technology.

ART + COM changes its goals in 1992. Greater emphasis is attached to software development. The change from Berlin to Bonn is therefore more difficult for us than the move to the "German National Research Center for Information Technology (GMD)" where Wolfgang Krüger is in charge of establishing a Department of Scientific Visualisation. He wishes to incorporate the artistic and scientific aspect that he has come to know through joint work in Berlin. As Fellows of the Academy of Media Art (KMH) in Cologne we are invited to work on our own projects. The base for this interdisciplinary work changes following Wolfgang Krüger's departure from GMD (1994) and his early death (1995). A physicist highly respected in the world of computer graphics, a friend with a keen interest in art, and a counsellor always ready to encourage and inspire others – such a person will always be sorely missed, but even more so in a technology-driven and industry-oriented environment such as the GMD National Research Centre with its some 1000 employees.

Nevertheless, award winning works were created over the years 1992 to 1997 in the fields of "man-machine communication", "unsharp interface" and "inter-

---

is put forward in 1988 by Monika Fleischmann (art educationalist and artist) and Wolfgang Strauss (architect) and is funded by the Berlin Senate for Science and Research.

[5] The project "Visualisierung digitaler medizinischer Daten" (Visualisation of digital medical data) is put forward in 1988 by Dr. Wolfgang Krüger (physicist) and is funded by the Berlin Senate for Science and Research.

[6] Eco U (1990) Über Spiegel und andere Phänomene. dtv, Munich

[7] Sonvilla-Weiss St (1996) Narziss, Selbstbespiegelung oder Selbsterkenntnis? Narzissmus im Spiegel der Gesellschaft. Federal Ministry for Education and Cultural Affairs, Medienservice, Vienna

active storytelling" and included "Responsive Workbench", "Liquid Views", "Rigid Waves", "Spatial Navigator" and "Skywriter". Productions such as "Video Only – Virtual Performance" or the award-winning Cyberstar[8] concept "Paramatrix" are applauded by experts worldwide. There is considerable international interest in these projects from the worlds of art, science and technology. In addition to art-oriented works dealing with identity and perception, new forms of communication are also examined in other research projects. There are plans to establish an independent area for "Media Art Research Studios MARS" in the Institute for Media Communication newly founded at the GMD. This is essential, however, to ensure continuous cooperation between art and science. Otherwise, work will tend to concentrate solely on the development of a "common language" on a project-specific basis. An Interactive Computer Installation researching new ways of communication can never be left to the individual. The production team can be compared with a large-scale team of architecture working on a competition project.

Interactive artistic work is a synthesis of theory, technology, design and play. Future work, learning, teaching and playing – the influence of digital technology on the image – the power of virtual reality – changing perceptions – sight and movement – the dynamic perspective – the dissolution of space. All these are areas that we keep coming back to, implementing them in architecture, art, dance and theatre. We have approached the task as artists and architects. Working closely with digital media and computer scientists we are turning into researchers and inventors. We see it as our job to bring poetry into media art and to address the visitor's senses. We are turning the theory on its head that man is losing his body to technology. In our opinion, the interactive media are supporting the multi-sensory mechanisms of the body and are thus extending man's space for play and action.

The Space as Stage Setting – a Game with Reality

Our work is based on recounting a story in space. It is unimportant in this regard whether the story is recognised. The only thing of importance is that the visitor finds a thread that stirs memories inside him. Three elements determine the structure – the spectator, the story and the space. The area of conflict is the relationship that exists between man and his world of experience. The immersion into a story which is presented virtually and interactively as a framework for human actions has a fundamentally different purpose than film or theatre. The basic form of the theatre is dialogue and contemplation of the various ways of thinking and behaving. Film is the modern form of dramatic art and represents a progression from the static image to the moving one. Interactive media art has the task of bringing the qualities of the theatre and film into confrontation and of intermeshing these qualities. The audience is given an unusual responsibility in this regard. They play a role in the story. And have to develop this role themselves in this game of illusion.

But what becomes of man in this staged world? How can he become part of

---

[8] The Cyberstar competition and the award-winning concept are described in: Fleischmann M (1995) Cyberstar 95 – Visionen zur Gestaltung der Zukunft. GMD Spiegel 3/95, GMD Sankt Augustin

it? What form of dialogue will he develop through interactive digital communication? How is the virtual space shaped by human intervention and what stories are told by man and space? We invite dancers to take up the new virtual play models. It is not just that the scenarios that are virtual, the settings themselves also change. Reacting stage backdrops and corresponding sounds generated by body movements give rise to crazy body gestures and choreographies. The unusual arrangement of audience and performers in space creates a state of confusion and uncertainty [9].

Commercial television productions such as magazine programmes or reviews of the year are performed in GMD's Virtual Studio by a number of German broadcast stations. Technical further developments match the Virtual Studio software to the practical needs of the market. At the same time, the European Commission is also promoting technical development as part of the ACTS programme. Unfortunately, however, there is no funding for developing flexible software geared to the needs of audience and performers. There is almost no support for the design and construction of virtual scenarios or for drama productions with their costly rehearsals. The broadcast covering the award of the 1995 Video Art Prize [10] can therefore only be realised through a technically oriented GMD research project (Distributed Video Production). For the first time ever, the Virtual Studio at GMD therefore undertakes a collaboration between dance and stage which takes place at two different locations. The virtual sets are transmitted live from Bonn via a ATM broadband to the studio stage in Baden-Baden.

Few Virtual Studio productions can lay claim to an image language that is adequate for the medium. In most cases, the individual image layers are not interwoven and presenters react dispassionately in the empty blue room. If they are indeed able to react to their visual environment and disappear through virtual doors, then they only succeed in doing so when working with a highly specialised team. The presenter is merely a puppet in this game. Be that as it may, the empty room is the interesting aspect in this scenario. To ensure that the actors do not get lost in this blue room, they must commit the scene to memory. They must really act and not just go through the motions. The emptiness is a challenge to their imagination, to a spatial mode of thinking and to reaffirming the body's gestures in space.

On 19 November 1995, some 150 invited guests in the Baden-Baden studio find themselves in front of and actually inside the virtual sets transmitted along

---

[9] We observe the public's confusion and the performers' curiosity during the television production using virtual sets which are produced live and at distributed locations on 19 November 1995 (similar to the presentation of the Video Art Prize in Schloss Birlinghoven and in Baden-Baden). Two days later, the two-hour live event is broadcast in a shortened, 40-minute version by WDR. The response is very gratifying, particularly as regards the design and staging of the two dance performances.

[10] The 1995 Video Art Prize is awarded by ZKM – Zentrum für Kunst und Medientechnologie and SWF – SüdWestFunk Baden-Baden. Jeffrey Shaw – head of the Institut für Bildmedien at ZKM – and Bernhard Foos – editor at SWF – commissioned us with the production of the entire real-life and virtual ceremony. With a small budget of DM 36,000, this experimental production could only be performed within the framework of a GMD research project (Distributed Video Production) and considerable investment of time and personal resources.

data lines the some 350 km from Bonn to the SüdWestFunk (SWF) studio. Two cameras, numerous actors and the audience are in blue box of the SWF TV studio, though the virtual sets themselves are actually in GMD's digital studio. The broadband network of Deutsche Telekom is used to transmit the camera shots of the actors to Bonn where they are inserted into the virtual sets and, without any noticeable delay, are retransmitted to Baden-Baden. For the first time ever, a blue box is used live as a stage. Eight virtual stage sets with their metamorphic build are used to give structure to the chronological sequence of events. The audience in the studio sees the actors in the virtual sets on the wall-high projection area – the "apparent" image that will be used later for the TV transmission – while the presenter and dancers live in the studio perform against the empty blue stage. As in the theatre, the performers appear from backstage. But here the voices appear from behind the audience's backs. The presenter appears in the image on the video wall. The scene of the action changes and also takes in the space occupied by the audience. The dancers leave the real stage through the audience area and apparently disappear in the video wall. The studio guests' attention is pulled to and fro. Part of the dramatic effect of the transmission is created by the fact that the audience are seated in swivel chairs, and thus create a similar picture to the movements of spectators on a tennis court. The game with reality remains the most important theme when working with virtual sets.

"Cyber City", Let Your Finger Do the Walking
"Berlin – Cyber City"[11] or How do I step into the virtual city? This study is the first of its kind to examine audience anticipation and the use of interactive systems in public spaces. The fall of the Berlin Wall provided the impetus in 1989 for us to take a closer look at our city. The reconstruction of the former capital now reinstated presents a major challenge in urban planning and one for which no-one is prepared. We are interested in making the various plans accessible to the public as a virtual reality game. But how can we convey the complexity of urban planning to a large audience?

The entrance to the "Cyber City" is an aerial shot of Berlin which is secured to a table and forms the reference level of the real city. We play the "let your finger do the walking" game and use an electronic thimble (Polhemus) to move around, show and visualise. The thimble is a sensory mechanism that conveys its positional data on an ongoing basis to the position detector secured underneath the table. The real location on the aerial photograph can thus be coordinated precisely in the computer with the 3D simulation of the city architecture. The visitor gains both an overview and an insight into the situation. The wall-high projection screen behind the table allows the visitors to follow their virtual trip through the "Tiergarten", past the Congress Hall (now the House of Culture) and the Reichstag. "And this is where we ought to be able to take a stroll through the Brandenburg Gate," calls out an enthusiastic East Berliner and is amazed when he finds he really can "drive through it, turn around and can then even fly back over it".

[11] Strauss W (1991) Cyber city flights. Leonardo – special, 10/91, Munich

The table is a metaphor for language and encounter which actually functions. At the international radio and television exhibition in Berlin in 1991 visitors are not discussing the new VR technology but rather what had happened in 1989 when the Wall fell. The virtual table turns urban planning into a discussion of the city that incorporates both past and future. The "Cyber City" can be compared in form with a video sculpture. Set up in a public space it consists of the two elements – a table and a video wall. There are only two main perceptual surfaces: the horizontal (the table with the overview plan that corresponds to the lie of the city) and the vertical (the large video wall which embodies the city facade). The observer becomes a stroller through a virtual film set.

Home of the Brain – the Computer's Memory

While the observer is only the onlooker, this "looking" is a kind of movement. It embodies "active observation". From a certain moment when the observer becomes immersed in the action, his "passive onlooking" is replaced by "active observation". The observer discovers that he – and not the artist – is the one creating the situation. When the situation changes and the observer becomes a player, he suddenly begins to identify himself with the situation. Observation becomes more than merely consumption. In this moment consumption ceases. This is all the more true in interactive scenarios when the observer participates in the game and can intervene in it. In 1990 we endeavour to construct Alice's Wonderland. With virtual reality goggles and gloves, the body is exposed to new spatial experiences. The body is the interface between the interior and the exterior, between reality and virtual reality. "Home of the Brain" [12] – depicted as a metaphor for the computer's memory – is awarded the Golden Nica of Ars Electronica in 1992. The work is a vision of the future of telecommunications. Four year later it will be possible to work with a similar version on the Internet. The Internet is already being used as a public forum, as a venue for the virtual representation of masks, avatares and agents (intelligent advisers). This vision was still Utopia back in 1992.

"Home of the Brain" is a three-dimensional mandala. Every visitor can move around in this virtual environment using the virtual reality glove and finger gestures. The performer's gestures will become immediately visible to himself and his audience through the representation of his hand. The entire production can be observed on monitors or a large video screen. The performer functions like a kind of shadow artist in the virtual space behind the screen. "The virtual hand discloses its true soul to us," explains neurologist Hinderk Emrich commenting on the virtual flight and lively movements of a physically handicapped participant in Geneva whom we are watching via ISDN lines from Berlin. Below the head mounted display he cannot see anything of the outside world and instead sees himself as an integral part of the new virtual world which surrounds him. For a short time he feels himself free of his real body. During the virtual flight he

---

[12] Fleischmann M, Strauss W (1992) Home of the brain. Golden Nica for interactive Art. Prix Ars Electronica, Ars Electronica 92 Catalogue, Linz

sets his own agenda and develops his own personal perspective of sound, since the objects are interactively associated with sounds, noises and fragments of text.

The "Home of the Brain" is inhabited in virtual terms by pioneers in media development. The thoughts of Vilém Flusser, Paul Virilio, Joseph Weizenbaum and Marvin Minsky are implemented in the computer's memory. "Do we need that? Why do we need it?". Weizenbaum's warnings against the power of the computer and the impotence of reason wrap themselves around his "House of Hope" on Moebius-like chains of thought. In Virilio's "House of Disaster", the "racing standstill" is tested under trees falling as if in slow motion. Flusser's "House of Adventure" shows his vision of flowing space: "I dream of a house with walls that can be changed at any time, of a world whose structure is no more than an expression of my ideas". In Minsky's "House of Utopia", a crystalline transformation object, future computer generations are discussed "which are so intelligent that we can be pleased if they keep us as pets". The "Home of the Brain" has anticipated paradigms that today are at the very heart of discussions relating to media communication. They include the organisation if information in virtual space, telepresence, information linking and interaction with objects in virtual space.

The Responsive Workbench – a Reactive Environment

Man is a mover. If man does not move, he is dead. We have learnt to move our "head" alone. The rest has to remain still. Our society has long since run up against a brick wall, since everything in our head is also turning. Be that as it may, we do not want to remain stuck in old systems. Do we really have to sit still at our work? We want to use our hands. We want to draw, build models and not just be keyboard operators. We want to see these models through the virtual camera. We want to let our eye take flight and spring across the wall of reality. Instead of drafting plans we want to produce 25 frames per second. Film language is exerting an influence on architecture. We are developing a photographic pattern of thinking.

In 1994 we design the "Responsive Workbench"[13] as a virtual work desk. The rigid arrangement of computer monitor and keyboard is to be replaced by a real training situation in which architects, engineers, medical staff and scientists can check and change their work in a simulated environment. The "Responsive Workbench" is a further development of the "table" metaphor used in the "Cyber City" project. Real-life situations and activities have been examined as to whether they can be transferred to virtual reality. The haptic checks with activities such as sketching, drawing, writing and painting are performed intuitively when we work with our hands. Kant calls the hand "man's external brain". The gestures of the hands and the gestures of speech control events on this reactive workbench. The person's own sight and body movements are connected to sensors that open up a dynamic perspective. The machine understands and reads our wishes for every possible observer standpoint and does so immediately from the eyes.

---

[13] Krüger W et al (1995) The responsive workbench: a virtual work environment. Virtual environments, Computer, IEEE

Fig. 1. People operating the Responsive Workbench. © GMD, VisWiz

Sensor-controlled stereo goggles makes the objects under the interactive glass projection table appear as transparent holographs. Virtual houses can be designed and changed with a data glove. Every angle of vision, each one of my body movements is recalculated in real time as a function of the virtual object. In medical simulations, the beating heart of a virtual patient can be lifted out, removed and examined from every angle. A self-learning speech recognition system reacts to specific commands in order to keep the hands free for other operations. The user interacts with the virtual scenario, displacing, changing and manipulating it in order to test it for realism. He can also retrieve information from the computer which works invisibly in the background. The objects and activities themselves become the inputs and outputs for this environment. These is no longer a clearly perceivable interface between the user and the system.

Virtual Reality and Interactivity as Medium – the Dissolution of Space

Painting, photography and television traditionally assume a static observer who, since the development of the frontal perspective in the Renaissance, has symbolised a distanced, quasi-objective approach. The technologies of virtual reality, on the other hand, anticipates a moving observer who himself is IN the image. Dynamism and constant change are the key features of interactive media, the illusion also encompassing the observer. His movement and location in space determine perspectives and the way of seeing things. He is IN the illusion. Linear spaces with static perspectives and fixed observer standpoints are thus history. Images are becoming virtual spaces unhindered by boundaries. The space is no longer a place, but rather a means.

In physical terms, the observer was always an outsider in the fictive worlds of cinema and television. His involvement in the course of events, in the fiction, called for emotional intelligence, identification and catharsis on the part of the fixed observer who was firmly planted in his seat. With interactive simulation techniques, on the other hand, it is not the mobility of the eye alone that is demanded but of the whole body. From these aspects, the technologies of virtual reality can be linked with other illusion technologies such as panoramas, relief cinema and stereoscopic photography which also enable the eye to move around at will. As with panoramas from the 19th century[14], interactive media allow us to develop new dramatic forms of storytelling. The dynamic approach of VR systems is replacing the static perspective of the Renaissance.

In the interactive VR environment, the image space is losing its fixed boundaries. At the same time, while the body's sensation is reinforced, a new feeling for spatial orientation needs to be developed. Identifying the position of the eyes, head and body – like the identification of gestures and speech – has the purpose of harnessing the human senses for directly controlling communication. Man must not be asked to change his body and senses to match the machine. Instead, the machine must to tuned to man's needs. To a far greater extent than with traditional media, the VR media interface serves as a key to the media work and thus determines both the dimension of interaction and the dimension of perception.

Skywriter – Navigation Through Body Balance

Like Hermes the celestial messenger, the observer navigates as a "Skywriter"[15] using "virtual balance" and the metamorphosis of digital landscapes. To do this, he uses neither mouse, joystick or data glove. He simply has to move his body's centre of gravity accordingly to allow him to fly upwards or downwards, to the right or to the left. Unlike a joystick or mouse which reduces man to minimal reflex actions, "Virtual Balance"[16] requires the coordinated use of the entire body and its perception. Neither time optimisation nor disjointed gestures are required, but rather an interplay of the senses. Apparently without effort, the "Skywriter" is able to fly through virtual landscapes. Linear storytelling is translated into interactive action and transformed into virtual space-time. The dramatic effect of the action is governed by the person's relationship to his own body. Here, too, we observe physically handicapped persons who are motivated in their movements. The ground below their feet becomes an interactive surface and the body's perceptual sensitivity coupled with body balance become a control instrument.

"Virtual Balance" is a navigational system for controlling images through the use of the body. It is also a platform for observing the effect of images on the body.

---

[14] Crary J (1990) Techniques of the observer; MIT, Massachusetts Institute of Technology, Cambridge

[15] Strauss W, Fleischmann M (1996) The role of design and the mediation of contents; Proceedings CADEX'96; IEEE, Computergraphics

[16] Fleischmann M et al, The virtual balance, Proceedings 6th Interfaces, Man-machine interaction, Montpellier '97

Fig. 2. Skywriter - Surfing the Webspace. © GMD, VisWiz

In the "Telepolis" 1995 exhibition in Luxembourg, Luxembourg's Grand Duchess accompanies her tour through the virtual city of Xanten with real-life jumps and reinforced body movements. During the presentation at CeBit '96 in Hanover, neurologist Hinderk Emrich finds himself repeatedly in dance situations and discovers there an "enthralling" perspective of the virtual world.

"Virtual Balance" was developed in 1995. It consists of a platform with 3 weight sensors and is controlled solely by the changes in the position of the human body's centre of gravity. Movements and gestures, the body and the entire perceptual apparatus become the interface. The observer's positional information is passed to the graphical system for the purpose of calculating the current image and the required information. Depending on the level of detail of the virtual model, which is calculated from the distance to the virtual objects, different information content is made available to the observer.

This first application is part of a global navigational concept that can be accessed via the Internet or as a permanent installation on site. The "Skywriter" will then fly through virtual continents, eavesdrop on the sounds of the various cultures, or discover the symbol sets of the different peoples. The "Global Passage" around the virtual world is intended to visualise cultural identity and convey this between different cultures. In the longer term, it will be possible to control this virtual world tour using two synchronous interfaces. The navigator will then be accompanied by a second "Skywriter". The coordinated movement of the two navigators is then used to control a shared virtual trip. The multimedia navigational environment is ideally suited for public spaces, banks, department stores and museums in order to be able to make contact with other cultures when travelling to another world or a different location. "Virtual Balance" is envisaged as an interface for navigating in the three dimensional net space, for surfing the Internet, for children, players and performers.

The Playable Instrument or the Sense of the Senses

A number of questions need to be answered when working with (interactive) media art. How can I create the instrument for playing with? How can I produce a work that is to be grasped through the senses alone, without any need for spoken language or written instructions? How can I make the observer understand what he can do? The observer should feel "it", that spiritual element of the work that is communicated through his body! What senses should be addressed? What cultural conditioning are to be incorporated or overcome?

The longing to penetrate a virtual space is something we are all familiar with from fairytales and films. Lewis Caroll's "Alice in Wonderland" trips head over heels down a rabbit hole and into another world. In Jean Cocteau's film "Orpheus in the Underworld", Jean Marais passes into the underworld through a mirror. He appears to penetrate the mirrored wall bereft of his physical body and finds himself in another time-space system. This is represented in the film by minimal body movements shown in slow motion. With the innocent eyes of a child, Alice discovers a world that grown-ups have long since forgotten. Her desire to play stimulates Alice to action. Her longing for adventure overcomes her fear of the unknown. Orpheus' love of Eurydice drives him to heroic deeds. Through their experiences in another world, Alice and Orpheus begin to see things in a different light and return to reality with a new outlook and zest for life.

In both cases it is specific symbols – the rabbit hole and the mirror – that mark the transition from one world to the other. The transition to another world is a key prerequisite for fairytales. Likewise the interface to the virtual world conveys different languages and perceptions. The interface is the key to the imagination. What is important is to push back the boundaries of perception and, wherever possible, to climb over these. In Cyberspace, the imagination space is a "house of illusion" that is generated in the machine and is projected by the body onto the body.

Rigid Waves – Approaching One's "Self"

"Rigid Waves"[17] transforms the acoustic mirroring of Narcissus and Echo into visual form. Narcissus gives up his body to his mirror image. The "self" becomes another (body). His own movements are only an illusionary echo. As the observer approaches the mirror, he is confronted with a mirror image that does not correspond to his normal perception of things. He sees himself as an impression, as a

[17] Fleischmann M, Bohn C, Strauss W, Rigid waves – liquid views; Visual proceedings: Machine culture; Siggraph '93, Anaheim, LA

body with strangely displaced movement sequences and, ultimately, as an image in the mirror that smashes as soon as he comes too close. He is unable to touch himself. A small camera hidden in the picture frame is used to place the observer in the image. The computer-controlled projection surface is controlled by an algorithm that calculates the distance to the observer. "Rigid Waves" is a virtual mirror which does not reflect but rather recognises. Sight and movement, approaching and distance are triggers for the unusual images. This is an attempt to see oneself from the outside, to stand side by side with oneself and to discover other, hidden "selfs". In this fractured mirror, we are able to find ourselves, our "self" has been liberated. But how will I ever recognise myself again?

Liquid Views – the Virtual Mirror

The central theme of "Liquid Views"[18] is the well in which Narcissus discovers his reflection. He initially sees water as someone else, as another body. Like the small child in the various "mirror stages" described by Lacan, he decides to recognise his fictive body as himself. This installation has the objective or arousing the observer's curiosity and seducing him to undertake actions that bring him into contact with his senses. There are no written instructions of the keys to be pressed – as is often the case with computer installations. Instead of pressing keys and buttons, the observer must experiment with his own sense of touch. What is difficult about this is that the visitor is normally prohibited from touching exhibition pieces. A disused underground station under the Madrid Opera House is therefore a far more suitable location for the exhibition than a traditional museum.

Attracted by the sounds of water and a room of shimmering lights, the visitor approach the virtual well. Seeing the image of himself he is tempted to touch it. Touching the image with his fingertip, the image in the water breaks up. Drawn by the sensation triggered by touching his own image in the water, the observer immerses himself in the situation. Liquid Views is also a metaphor for the Internet: while the person becomes lost in his own actions, he leaves traces behind and is monitored.

The images of the visitor are stored in the computer. During subsequent analysis we can spot the differences in communication behaviour of different cultural groups. Apparently, there is nothing more true than the cliché of stereotypical behaviour shaped by culture. Most people in Bonn are sceptical. In Zurich most people appreciate the installation for its aesthetic value. While in Hollywood we generally encounter artists who want to include our installation in their video clip, film or bedroom. "This is like having sex with my computer", whispers Coco Conn.

The two works "Liquid Views" and "Rigid Waves" break entirely new ground in the field of unsharp interfaces and virtual reality. Visitors find themselves wanting to touch the surface of the water or to alter things by changing the mirror image of themselves. These are reactions which make sense. The interface is not inter-

[18] Fleischmann M, Strauss W (1996) Internet – a digital muse? In: Media Art Perspectives. Edition ZKM, Cantz Verlag, Ostfildern

Fig. 3. Liqiud Views. © GMD, Monika Fleischmann, Wolfgang Strauss, Christian Bohn

preted as such. It goes unnoticed and is not consciously perceived. These natural references turn Liquid Views and Rigid Waves into virtual reality. The interactive behaviour of the observer as it is observed in the exhibition process is an integral part of the work and will find its way into new concepts.

The art educationalist Christoph Liesendahl gives his impression of the two works:

"While the water in "Liquid Views" seduces the observer and draws him into its depths, visitors to the "Rigid Waves" work are afraid of the faceted image they see before them. Both works confront the observer with himself in different ways and examine how we react to our quickly changing surroundings. The body becomes an interface to a spatial experience in a virtual reality where it can itself determine how things are observed and the speed of the spatial experience itself. In doing so, it learns to stem off the flood of data and yet to play a role in the world of the telepresence.

Typically, the presence of space in coordinating one's own interaction plays a key role in both works. Both works look like environmental video installations. As "Liquid Views" is presented to the public in exhibition halls worldwide, the image space of the water is projected onto the four walls and the floor of the exhibition area. While the interactive observers forget their real surroundings and 'sink' into this virtual space, the onlookers are immersed in the water and its sounds and find themselves an integral part of this interactive environmental installation."[19]

The Interface – an Invitation to Communication

Learning to see through the slowness of movement. Learning to hear through long forgotten sounds uninfluenced by images. Feeling what isn't there, in order to understand the synthesis. The poetry of the interface determines the dimension of interaction and one's own experience. What am I doing here? What am I experiencing, feeling? Stroking my hand over the water, coming face to face with myself, flying through space, holding my hand over my eyes. What does it all mean? The visitor is invited to examine these questions. It is an invitation to communication.

Man has lost the use of his senses. Eye/gesture/movement/speech recognition or body balance is used to test out the body as an interface. We attempt to rediscover the senses through the use of technology. In the virtual space, we practice for reality and live with a feeling of "as if". As if we are dreaming, as if we are grasping hold of the water, as if we are flying through the air, using just our bodies and the appropriate high-end technology. As if we are dying, falling, sliding, going into orbit, as if we are existing. We create "Imagination Systems"[20] – an electronic

---

[19] Liesendahl Ch (1996) Bildende Kunst der Gegenwart und technische Innovation. Zur Rezeption neuer Interaktionsformen von Körpern im Raum. Diss. Mag. Art. in Philology and Art, Johann Wolfgang Goethe University, Frankfurt am Main

[20] Fleischmann M (1995) Imagination systems – Interface design for science, Art and Education; Proceedings IMAGINA '95, Monte Carlo

aura for ourselves. And do all this with the objective of existing in this or another reality.

There are still too few institutes that are examining these questions. Independent research sections are needed for media art and culture in order to be able to work continuously. Creativity and fantasy are the raw materials of the 21st century. Interdisciplinary teams can examine global problems and carry out artistic and scientific experiments. Cooperation between technology and industry is currently seen as a crucial measure for making large-scale research centres more flexible. Cooperation with theatre, dance, film, television and museums would lead to new products and productions, which would also show progress-driven industrial partners the need for reflection and the "ethics of preservation" (Hans Jonas).

Unlike "virtual" real-time, the real-life "now" time requires rapid action. A draft for the society of the next century cannot exclude today's communication instruments. Individual actions and collective responsibility must be today's goal – and not the mentality of constant progress. The time has come to test out the world of possibilities offered by virtual reality in order that we can remain functional in reality itself.

Art as a Living System

Christa Sommerer and Laurent Mignonneau

"I don't think you have to worry about the future.
It isn't here yet. And now it is gone."
John Cage

1. Introduction and Conceptual Background

From Newton's mechanistic world view of the 17th century, which attempted the exact and mathematical description of nature and thus shaped our perception of "the world as a machine," to the Cartesian division of mind and matter, which still significantly influences our thinking today, we can observe a slow but steady shift toward a more abstract and diversified notion of nature in the late 20th century: "Natural science is not any more the spectator of nature, but realizes itself to be an integral part of the interplay between human observer and nature."[1] As Bohr continues, "... we now come to understand, that we are not any more the audience, but always also members of the cast of life."

At the end of the 20th century this thought is taken up again, as Gaston Bachelard[2] expresses it " .. from now on Hypothesis means Synthesis."

The universe according to Bohr[3] is an indivisible dynamic whole who's parts are essentially interrelated on a subatomic level. Fundamental is the dual character of the subatomic particles; they are two complementary descriptions of reality, where both are partly true and confirm the "Uncertainty Principle" of Werner Heisenberg.

From the insight that interaction itself and the interrelation between entities are the driving forces behind the structures of life, Sommerer and Mignonneau as artists investigate interaction and the creative process itself. Creation is no longer understood as expression of the artists inner creativity or "ingenium" (according to Hegel) but becomes itself an intrinsically dynamic process that represents the interaction between the human observer, his/her consciousness and the evolutionary dynamic and complex image processes of the works ("Art as a Living System"). Sommerer and Mignonneau assume, similar to Gregory Bateson[4], that the patterns of mind (consciousness) and the patterns of matter are reflections of one another and part of an unbroken dynamic whole.

2. "Art as a Process": Marcel Duchamp, Allan Kaprow and John Cage

"It is through Marcel Duchamp that we know that it is the observer, or the person who pays attention, who finishes the work of art. So nothing is done until each of us does it. And then it is done in a very unique way."[5]

Marcel Duchamp was the first artist to completely abandon the traditional art object by replacing it with the found object, the so called "ready made". In doing this, he substituted the artwork with the reflection of the object created in the audience's minds while perceiving a common object displayed as an art object.

John Cage is credited as the first artist to use random chance procedures and processes. Influenced by Zen Buddhism, he already in the 50's taught that ".. consciousness is not a thing but a process, that art must entail the random, indeterminate and chance aspects of nature and culture, that behavioral processes continually inform a work of art as an objective state or completed thing, and that "the real world .. becomes .. not an object [but] a process."[6] In an interview conducted in 1989 he states that "So our minds change with regard to the answers to our questions. We discover that all the answers are good, rather than there being only one good one and the rest being wrong, which of course was a nineteenth century idea."[7]

Integrating the audience into the art process has been explored by artists of the Fluxus and Happening movements as well as performance artists since the 60's and 70's. Artist Allan Kaprow, who invented the term "Happening," stated that "Casting into question boundaries between discrete art objects and everyday events and actions, the happening gave visual definition to the interstice between art and life." In his "Untitled guidelines for Happenings" of 1965 he wrote: "The guideline between art and life should be kept as fluid, and perhaps indistinct, as possible."[8]

3. "Interactivity" as Art Form

With the advent of new computer technologies new dimensions were introduced to the artistic creation process: virtual space and real-time interaction.

Advances in computer technology presented a new medium to artists. A group of artists centered around the Städelschule Institute for New Media in Frankfurt, Germany elevated interaction to one of their primary goals in the creation and presentation process.

Interaction was defined as human-machine and human-computer interaction to facilitate audience participation. A new medium called "interactive computer installations" was created. Artists such as Peter Weibel, Jeffrey Shaw, Christa Sommerer, Laurent Mignonneau, Ulrike Gabriel, Michael Saup, Agnes Hegedüs, Christian Möller and Akke Wagenaar exclusively used computers and interfacing techniques to create audience participatory interactive art works. New methods for "interaction" and "audience participation" were being explored, investigated and tested in their various forms: mouse, keyboard, data gloves, data helmets, pressure sensors, Polhemus sensors, midi interfaces, sensor platforms, joysticks and other devices were tested and connected to the image or sound worlds.

4. Multi-layered, Non-linear Interaction

While most of the above artists applied the various interface devices, Sommerer and Mignonneau became increasingly interested in non-deterministic, multi-layered and non-linear interaction.

When they started to create interactive art works in 1992, most existing interactive works followed a predetermined path of interaction: artistic CD ROM's and interactive art works were designed by authors to give the users a variety of different choices and paths to follow, but the discovery of unexpected new paths of interaction were rather limited.

Sommerer and Mignonneau believe that interaction is interesting to the user if, rather than being linear or predictable, it feels like a journey. The more one engages in interaction, the more one learns about it and explores it. Non-linear and multi-layered interaction should be easy to understand from the very beginning and should open pathways to continuously evolving levels of experience.

5. Natural Interfaces

Sommerer and Mignonneau have been particularly interested in the development of so-called natural interfaces, as they convey the consciousness of life, variation and personality.

In our daily lives we find ourselves continuously confronted with a complex ever-changing environment in which we have an inchoate understanding of the relationships between different objects, persons and events. However, we can cope with such bewildering situations by managing to adapt ourselves through learning, curiosity and the process of filtering knowledge for what is truly important.

Natural interfaces are intended to function in a similar way, by making use of complex interactions, adaptations and learning.

For example, adopting living plants as an interface not only provides a new and unusual connection between computers and a living being but also poses the questions of what a plant is, how we perceive it, and how we can interact with it.

Natural interfaces allow us to project our personality into virtual space. They also avoid the uneasiness of putting on unpleasant devices for entering virtual space (i.e., unencumbered interaction). Natural interfaces have included living plants, water, light and camera detection systems.

6. Process-oriented Art versus Object-oriented Art

In 1992 Sommerer and Mignonneau developed the concept of natural interfaces and evolutionary image processes linked to interaction.

They started working with evolutionary biology and became increasingly intrigued by how natural evolution and the processes of nature can function as a tool of creation.

They adopted evolutionary image processes to create process-oriented art, rather than pre-designed, predictable and object-oriented art. In art, much of the production is centered around the art object, the artifact; even in interactive art many artists still heed the traditional notion of the art object.

When we analyze the very essence of digital technology, we realize that it is the capability of creating, rendering and displaying processes that distinguishes

Turing machines[9] and computers from other media such as photography, film and video.

The potential rewards of developing, emerging and evolving processes became a focal interest of the authors artistic investigations. Instead of presenting the audience with hand crafted artifacts or art objects, they pioneered process-oriented art works.

7. "Interactive Plant Growing": a Novel Way of Interaction

The first interactive computer installation that did not use such devices as joysticks, mouse, data helmets of trackers was called "Interactive Plant Growing."[10] Real living plants functioned as an interface between the human visitor and the art work. This work was later praised as "Epoche making" by Toshiharu Itoh, ICC 1995[11] not only for its use of a new and unusual interface but also for its concept of interaction with a living being and the transformation and interpretation of this dialogue in a virtual environment.

Fig. 1. Interactive Plant Growing – installation view. © 1992, Christa Sommerer and Laurent Mignonneau, collection of the Media Museum of the ZKM Karlsruhe

"Interactive Plant Growing", created in 1993, deals with the sensitive interaction between five actual plants and five or more human viewers who can, by moving their hands toward the real plants, initiate and control three-dimensional real-time growth of artificial plants. By engaging in a sensitive interaction with the real plants, the viewers become part of the installation: they influence how this human-plant communication is translated into the virtual growing on the computer display. The electrical potential differences between the viewers body and the real plants is interpreted as electrical signals: they determine how the virtual three-dimensional plants will develop. By touching or merely approaching the real plants the viewer engages in a dialog with the real plants in the installations (Fig. 1). He or she can stop, continue, deform and rotate the virtual plants, as well as develop new forms of plants and unexpected new combinations. As the growing processes are programmed to be very flexible and not predetermined, the result on the screen is always new and different, depending on the viewer – plant interaction.

"Interactive Plant Growing" was the first interactive installation where the visitors became essential to the development of the piece: without their interaction the piece could not exist, and images disappeared as soon as the visitors left. It was also unique in the sense, that minute personality and interaction differences could be interpreted in the form of complex image sceneries that solely depend on each viewer's identity.

8. Artificial Life and Art: an Approach

Through their backgrounds in biology, Sommerer and Mignonneau became increasingly inquisitive about the process of creation itself.

'Artificial Life' (A-Life), a research field developed by scientist Christopher Langton at the Santa Fe Institute[12], proved capable of producing processes of nature within a machine (computer environment) and allowed computer programs to evolve over time. This enabled the development of processes and patterns that are no longer predictable or "hand made."

Fascinated by the idea of creation through evolution, not understood as a scientific simulation or mimicry of nature but as an investigation into the creative process itself, Sommerer and Mignonneau studied the possibilities of applying artificial life principles to art projects. [13]

Natural evolution has brought about a vast variety of forms and structures in nature: it became apparent that artificial evolution could function as a new mechanism of the visual creation process. Similar to John Cage's use of chance procedures in his musical compositions, Sommerer and Mignonneau began to implement a combination of interaction and artificial evolution to the creation process in their works.

9. "A-Volve"

They started working with A-Life scientist and creator of the "Tierra" system, Dr. Thomas S. Ray[13]. In their collaboration in 1994 they developed "A-Volve"[14], an interactive computer installation that allows visitors to create artificial creatures. These artificial creatures inhabit a water filled glass pool, where people can interact with them through touch (Fig. 2).

Artificial life principles are applied to the birth, creation, reproduction and evolution of these artificial life-like beings. The creatures are products of evolutionary rules and influenced by human creation and interaction.

By designing any kind of shape and profile with their fingers on a touch screen, visitors will "give birth" to virtual three-dimensional creatures that are automatically "alive" and swim in the real water of the pool.

Fig. 2. A-Volve – installation view. © 1994, Christa Sommerer and Laurent Mignonneau, supported by ICC-NTT Japan and NCSA, Urbana IL, USA

The movement and behavior of the virtual creature is decided by its form, that is, how the viewer designed it on the touch screen.

Behavior in space is, so to speak, an expression of form.

Form is an expression of adaptation to the environment.

Form and movement are closely connected, so the creatures capability to move will decide its fitness in the pool.

The fittest creature will survive longest and will be able to mate and reproduce. The creatures will compete by trying to get as much energy as possible. Thus predator creatures will hunt for prey creatures, trying to kill and eat them.

The creatures also interact with the visitors by reacting to their hand movements in the water. If a visitor tries to catch a creature, it will try to flee or stay still if it gets caught. The visitor is thus able to influence the evolution by, for example, protecting prey against predators.

If two strong creatures meet, they can create an offspring and a new creature can be born. It carries the genetic code of its parents.

Mutation and cross-over provide a nature-like reproduction mechanism that follows the genetic rules of Mendel. This newly born offspring will now also react and live in the pool, interacting with visitors and other creatures.

None of the creatures is predetermined: they are all born exclusively in real time through the interaction of the visitors and the mating process of the creatures themselves.

As the genetic code of the offspring is passed from generation to generation and the propagation of the system is based on selection of the fittest creatures, the system is able to evolve over time toward fitter populations. Although the evolution could take place by itself without outside influence, the system is designed so that the visitor's interaction and creation of forms will significantly influence the evolutionary process.

We can consider the visitors an external selection mechanism.

The three main internal parameters, fitness, energy and life span, regulate the interaction, reproduction and evolution among the creatures.

The external parameters are the visitors drawings on the touch screen as well as their interaction with the creatures (further readings about "A-Volve"[15]).

Because the social interaction between the viewers and the virtual world is essential for the creation of the work, we might call "A-Volve" a complex system where, as in Quantum Physics, the entities transform their states according to probability patterns. The system is like an interconnected web: intrinsically dynamic through its movements, interactions and transformations of particles and entities.

10. Entering Virtual Space

Driven by the desire to enter virtual space and to interact with virtual worlds in a direct fashion, Sommerer and Mignonneau in 1995 created and patented a new interface, the "3-D Video Key."

This system allows visitors to enter a virtual space totally unencumbered and to see themselves in the virtual space that surrounds them. They are able to cross the virtual space freely and in real-time to experience the virtual environment in a very natural way.

"Trans Plant"

In 1995 Sommerer and Mignonneau were invited to develop a permanent interactive computer installation for the Tokyo Metropolitan Museum of Photography as part of the museum's collection.[16] They created "Trans Plant" with the support of ATR Advanced Telecommunications Research Laboratories, Japan.

In "Trans Plant," visitors enter a room shaped in a semi-circle and immediately become part of a virtual jungle that starts to surround them. As the visitor steps forward into the installation space, he sees himself inside a projection screen in front of him (Fig. 3). By walking freely he soon discovers that grass is growing wherever he walks, following each step he takes and each movement he makes. When stopping and standing still, trees and bushes grow where the visitor stands. Changing the speed and frequency of his movements, he can create a biotope that is laden with different plant species. Size, color and shape of these plants depend on the size of the person. As each visitor creates different plants and his own personal environment, the experience becomes an expression of his personal understanding for the virtual space.

Fig. 3. Trans Plant – installation view. © 1995, Christa Sommerer and Laurent Mignonneau, collection of Tokyo Metropolitan Museum of Photography

With the same technical configuration used in "Trans Plant", Sommerer and Mignonneau created a second interactive installation called "Intro Act".

"Intro Act"
In 1996 they were invited to develop an interactive computer installation for the Biennale de Lyon as part of the museum collection of the Musée d'Art Contemporain in Lyon, France.[17]

In "Intro Act," visitors enter the installation space and immediately see themselves integrated into a virtual world. As they move their bodies in space, different three-dimensional evolutions of abstract organic forms are synchronized and linked to their movements.

As if exploring a different universe, a visitor will try to investigate which movement causes which image event. For example lifting an arm will lead to sudden explosive growths out of the hand.

Other behavior and movement can lead to destruction of organisms, whereas certain behavior will cause construction, expansion and differentiation of the virtual species. The visitor becomes totally engulfed in this virtual world, and the longer he interacts the more he becomes a part of the system. He observes himself inside the three-dimensional world, defines it, creates it, destroys it and explores it (Fig. 4).

11. The Building Blocks of Life
In the course of their studies of evolutionary image processes, Sommerer and Mignonneau in 1996 began to investigate the building blocks of visual creation and explored how simple structures can create complex shapes and forms through

Fig. 4. Intro Act – installation view. © 1996, Christa Sommerer and Laurent Mignonneau, permanent collection of Musée d'Art Contemporain, Lyon

genetic manipulation. They produced "GENMA – Genetic Manipulator," an interactive installation that permits visitors to create, manipulate and explore the genetic design of artificial creatures. "GENMA" was developed in 1996 for the AEC Ars Electronica Center in Linz Austria as part of its permanent exhibition.[18]

"GENMA" is a kind of "dream machine"[19] that allows manipulation of artificial nature on a micro scale. Principles of genetic programming are used to construct the creatures.

When looking into a mirrored glass box, the visitor sees those creatures as stereo projections in front of him.

Each creature's genetic code is schematically displayed on a touch screen. With his finger the visitor can manipulate the genetic code on the touch panel and thus alter and modify in real time its appearance in the glass box.

By selecting and merging different parts of the genetic string and then mutating and recombining them, the visitor can learn how to create complex forms out of seemingly simple structures at the very beginning.

On a visual level, "GENMA" explores the concept of "natural design" or "auto design," which is no longer prefixed and controlled by the artists but represents the degree of personal interest and interaction of the visitor. Each visitor shapes the forms he wants to see, aided by artificial genetics, mutation and his selection criteria.

12. "Life Spacies" – from Text to Form on the Web

In 1997 Sommerer and Mignonneau were asked to develop a new interactive installation called "Life Spacies"[20] for the permanent collection of the ICC Inter-Communication Museum in Tokyo, Japan.

"Life Spacies" is an evolutionary communication and interaction environment that allows remotely located visitors to interact with each other in a shared virtual environment. Visitors can integrate themselves into a three-dimensional complex virtual world of artificial life organisms that react to their body movement, motion and gestures. These artificial beings also communicate with each other as well as with part of an artificial universe, where real and artificial life are closely interrelated through interaction and exchange.

Through the "Life Spacies" web page (Fig. 5), people all over the world can

Fig. 5. Life Spacies – view of Life Spacies Home Page. © 1997, Christa Sommerer and Laurent Mignonneau, permanent collection of ICC Intercommunication Museum, Tokyo

contribute to the system by simply typing and sending an email message to the "Life Spacies" web site (http://www.ntticc.or.jp/~lifespacies) to create one's own artificial form.

These creature will immediately start to live in the "Life Spacies" environment at the ICC museum and interact with the visitors on-site.

After sending the email message, people will receive a small curriculum vitae about their creature, as well as an image of how it looks.

The artificial species can be created in two different ways:

a) through incoming international email messages. A text-to-form editor creates the genetic code for each creature:
   - one message is one creature
   - complex text messages create complex creatures
   - different levels of complexity within the text represent different species

b) by the creatures themselves:
   - reproduction helps the creatures to propagate their genotype in the system so they can form groups of different species

After sending an email message to the "Life Spacies" web page, the sender soon receives a curriculum vitae for his creature, as well as an image of how it looks. When the creature dies, a report is given to it's creator, telling him or her how long the creature lived and how many children and clones it produced.

"Life Spacies" is again based on the idea of evolutionary design, which is not predetermined by the artist but solely depends on the interaction of the visitors and the evolutionary process. Only the messages mailed from people all over the world and the reproduction and evolution of the creatures themselves will determine how the creatures will look and how they will behave.

One can thus not really predict how the work will evolve and what kind of creatures will emerge. It will exclusively depend on how many people send messages, how complex these messages are, and how the creatures reproduce among themselves and through the selection of the visitors in the museum.

In "Life Spacies" interaction, interrelation and exchange happens on human-human, human-creature, creature-creature, human-environment, creature-environment and life-artificial life levels.

As the interaction rules are non-deterministic and multi-layered, an open system was created where each entity, whether real life or artificial life, whether actually present (at the ICC Museum) or virtually present (the users on the net, or the creatures as code), is regarded as an equally important component of a complex life-like system.

13. "Art as a Living System"

In the various examples above, we have witnessed that the art work on Sommerer and Mignonneau's terms is no longer a static object or a pre-defined multiple choice interaction but has become a process-like living system ("Art as a Living System"). The art work is characterized by complex interrelations and interactions of real and virtual entities that engage in a dialogue and cause and effect the appearance of different expressions of mind and matter.

Relativity Theory proved that the cosmic web is alive by showing that its activity is the very essence of being. On an abstract level, the activity of these interactive systems could be considered alive as they are processes of continuous change, adaptation and evolution.

14. The New Position of the Artist

As pointed out earlier, integrating the audience in the art process has already been explored by artists in the Fluxus and Happening movements as well as performance artists since the 60's and 70's.

Also, other artists such as Wolf Vostell, the founder of Fluxus, and Ben Vautier overcame the traditional notion of the art object by replacing it through events and processes. Nevertheless, their ultimate aspiration was to create a work of art that was strongly driven, conceived and linked to the artist's personality.

While performance art often challenged moral, social and ethical principles, for example the extreme events of the "Wiener Aktionismus" or the performances by Yoko Ono or Marina Abramovic's and Ulay's "liberation" performances, they all seem to be still strongly preserving the boundary between artist and audience.

Around 1966 On Kawara, a Japanese born artist, introduced the concept of life and art as a process. Painting every single day between 1966 to 1985, over a period of 23 years, the same succinct image that only displayed the date of that particular day on a gray background, his paintings and the process involved could be described as an art work in progress[21]. Involving no technology besides his paint brush, his painting process could be considered a forerunner of the "art work as a living process." Furthermore, Vito Acconci can be credited as one of the first artists from the Happening and Performance movements to realize the potential of the new viewer-artist relationship, which is now being scrutinized in computer aided "Interactive Art." "To get back to "art," I have to make contact with those people who share in an art context: my space and viewer's space should come together, coincide."[22]

Sommerer and Mignonneau push the concept of "art as a process" a step further by not only relinquishing the "object of art" but also by skeptically questioning the position and function of the artist. They intentionally replace themselves through design processes of nature, such as evolution, selection and the complexity of virtual life in combination with the selection decisions of the audience. Doing this they consider themselves in the tradition of artists like John Cage by sharing his interest in the exploration of a world driven by behavioral processes rather than by objects.

Through blurring the border between life and art, they intend to develop the structure of the work through the interrelations among the audience members personal relationships to the developing and continuously changing inherent processes of the work.

This implies placing high expectations on the public: only if the visitor agrees to become part of the system, will he comprehend that there are no pre-defined solutions to be found within the art work but that instead he himself essentially determines what he sees. Each visitor will hence create his own artwork that is essentially a reflection of his inner expression and expectations. To push this idea further one might say that the visitor becomes part of the result and thus can be defined as a part of the art work.

In fundamentally rejecting the old nineteenth century notion of the "art object" and giving up the control over the art work, the artist indeed now provides a novel concept, where the usual art object is replaced by processes of nature or artificial nature and the artists persona is being replaced by the interaction between the audience and the art work.

15. Conclusions

Interactivity and the integration of life-related processes provide a novel concept in art as they empower the integration of personality, variety and a new viewer – art work dialogue. The artist who creates these installations only conceives the frame work: it is the visitors who form the art work through their interaction with each other, with the system and the image processes of the work. The results in these installations are not static, pre-defined and predictable but rather become like traces of living processes themselves.

Furthermore, the art work no longer confronts us with art objects but instead with art processes that are closely linked to the processes of creation itself.

References

1 Heisenberg W (1984) Das Naturbild der heutigen Physik. In: Blum W, Dürr H-P, Rechenberg H (eds) Werner Heisenberg: Gesammelte Werke, Vol. 1, München, pp 398–420
2 Bachelard G (1988) Der neue wissenschaftliche Geist, Suhrkamp, Frankfurt, p 12 ff.
3 Capra F (1979) The turning point, Bantam, New York, p 78 ff.
4 Bateson G (1982) Mind and nature, Dutton, New York
5 Cage J (1992) In: Tisdall C (ed) Art meets science and spirituality in a changing economy, University of Washington Press
6 Cage J, Charles D (1981) For the birds: John Cage in conversation with Daniel Charles, Boyars, Boston
7 See Ref. 5
8 Kaprow A (1996) In: Stiles K (ed) Contemporary art – a source book of artists' Writings, University of California Press, Berkeley and Los Angeles, p 682 ff.
9 Turing A (1950) Computing machinery and intelligence. In: Mind, 59
10 Sommerer C, Mignonneau L (1993) Interactive plant growing. In: Gerbel K, Weibel P (eds) Ars Electronica 93 – Genetic art artificial life, PVS Verleger, Vienna, pp 408–414.
11 Itoh T (1994) Approach to life – The world of Christa & Laurent. In: Christa Sommerer and Laurent Mignonneau catalogue, Tokyo: ICC-NTT InterCommunication
12 Santa Fe Institute, New Mexico, USA: http://www.santafe.edu
13 Kusahara M, Sommerer C, Mignonneau L (1996) Art as living system. In: Systems, control and information, Vol. 40, No. 8, Tokyo, pp 16–23
14 Ray T (1991) An approach to the synthesis of life. In: Langton C et al (eds) Artificial Life II, Addison Wesley, Redwood City, pp 371–408.

15 Sommerer C, Mignonneau L (1997) A-Volve – an evolutionary artificial life environment. In: Langton C, Shimohara K (eds) Artificial Life V, MIT Press, Boston, pp 167–175
16 Sommerer C, Mignonneau L (1995) Trans Plant. In: Moriyama T (ed) Imagination catalogue, Tokyo Metropolitan Museum of Photography, Tokyo, Chapter 2 ff
17 Sommerer C, Mignonneau L (1995) Intro Act. In: 3e Biennale d'Art Contemporain de Lyon, Réunion des Musées Nationaux, Paris, pp 378–381
18 Sommerer C, Mignonneau L (1996) GENMA-Genetic Manipulator. In: Stocker G, Schöpf C (eds) Ars Electronica' 96, Memesis the future of evolution, Springer, Wien New York, pp 294–295
19 Fuchs M (1996) Para real. In: Felderer B (ed) Wunschmaschine Welterfindung, Springer, Wien New York, p 212
20 Sommerer C, Mignonneau L (1997) Life Spacies. In: ICC Concept Book, NTT-ICC, pp 96–101
21 Kawara O (1988) In: Selected works from the collection of Nagoya City Art Museum, Nagoya City Art Museum
22 Acconci V (1996) In: Stiles K (ed) Contemporary art – a source book of artists' writings, University of California Press, Berkeley and Los Angeles, p 764 ff

Convergence of Art, Science and Technology?

Jeffrey Shaw

When one considers the visions which accompanied the development of contemporary art over the last 30 years – the kinetic art of the 1950s, the "open artwork" (happenings, environments, performances, land art, etc.) of the 1960s, the conceptual and social art forms of the 1970s – one finds that these visions have interesting and astonishing parallels in the technological developments of the 1990's. Interactivity creates an intimate relation between the artwork and the viewer, telecommunications permits a radical extension social interactions, simulation gives direct form to conceptual propositions. Of course we cannot foresee if science and technology will bring about a fulfillment or finale of these utopian artistic movements. But certainly there is an awakened desire to embody ourselves within these new territories of endeavour and experience.

The activity off both art and science has always been the interpretation and recreation of reality. It is an exercise of the human imagination, creating virtual realities which embody tentative structures of meaning. The world appears to us in the light of these fictions that we project onto its surface and art arbitrates this discourse between reality and illusion. The traditional activity of art has been the representation of reality – the manipulation of materials to create tangible mirrors of our experience and desire. Today the simulational effectiveness of the multimedia and televirtual technologies offers us a new medium of expression and also the cosmography of a new space of visions and visualization. Using the mechanisms of the digital technologies the artwork can become an immaterial digital structure encompassing synthetic spaces which we can literally enter. Here the viewers are no longer consumer in a mausoleum of objects, rather they are travelers and discoverers in a latent space of sensual information, whose aesthetics are embodied both in the coordinates of its immaterial form and in the scenarios of its interactively manifest form. In this temporal dimension the interactive artwork is each time re-structured and re-embodied by the activity of its viewers.

Certain characteristics of the new technologies are significant. Interactive computer graphics has become a shared language in many fields of research, and as a consequence a great diversity of information coexists that can be correlated in the digital environment. This is a unique situation historically and culturally, one which artists and scientists can take advantage of to forge a new discourse. One consequence is that art and media technology centers are being established worldwide to facilitate a new meeting ground between artists and scientists. And the proliferation of media events, exhibitions and festivals also presents pragmatic evidence that art and science are experiencing a vigorous renaissance of creative interrelationships.

Multimedia telecommunication and networking is heralding a fundamental transformation of our social and cultural paradigms. Tele-virtual-reality has become the appropriate domain for expressing our technological and artistic desires. The

Fig. 1. Legible City. © 1989–1991 Jeffrey Shaw, Collection of the ZKM Mediamuseum Karlsruhe

new modalities of interactivity, simulation and virtual reality are able to configure an immaterial yet tangible "newfoundland" of forms and images which we can enter and explore. This fictitious cosmography is searching to constitute those spaces and forms in which it can fully manifest its imaginative orbit.

Of great importance is the research and development of new codes and mechanisms of spatial representation. This has been a basic preoccupation throughout the history of western art. The formulation of a set of spatial coordinates, in for instance Renaissance perceptive, provides an underlying aesthetic and existential paradigm within which a culture achieves tentative representation, and thus comprehension of its desires. The recently developed digital imaging technologies offer the artist new methods and new paradigms which extend the spatial identity of the artwork. And not just in terms of the structure of the image itself, but also in terms of a space of interaction between the image and the spectator. On the level of representation, dynamic spaces can be built with inverted perspective, impossible architectures and infinite amplitudes. In their interactivity these works create bridges into the real environment of the viewer, conjoining virtuality and actuality into a coactive space of dramatized aesthetic experience.

It is also significant that virtual reality is so often an activity of world building, of the creation of socio-urban meta-architectures. Following from the Situationists we recognize that the city is simultaneously a tangible arrangement of forms and an immaterial pattern of experiences. Its underlying identity is a psychogeographic network of information – a labyrinth of narratives secreted within its urban framework. The new technologies allow us to create mediated cities that mirror the objective world into this virtual imaginative space. Simulation deconstructs a city's weighty material structures and evokes a fluid poetics of tentative space.

Fig. 2. Virtual Museum. © 1991 Jeffrey Shaw, Collection of the ZKM Museum für Neue Kunst Karlsruhe

An important paradigm is the "virtual museum". This is not a museum in the traditional sense but more like a "memory theater" where an interactive meta-architecture embodies a store of audio visual information in a form that hybridizes the functionality of a museum, a library and a game arcade. Furthermore the extension of such virtual museums into the network creates a ubiquitous and universal space of access to informational and artistic structures. The nowadays practice of art is bound to traditional structures of exhibition, publication, consumption and economics. Worldwide digitally networked communications offer the opportunity for the development of completely new forms of propagation and dissemination of creative activity. Space, time and interaction become the design parameters of the televirtual ambiance, and the mass address of the televirtual ether can take the practice of art from the periphery into the center of all social discourse.

One of the pertinent issues in this immaterial cyberspace of forms and ideas is the telepresent extension of our bodies through space and time that these technologies afford us. The technological deconstruction and artistic reconstruction of our identities in the digital ether is an almost meta-physical enterprise. One can say this despite the apparent ludic simplicity of the games we create (on the Internet for instance) to embody this surreal multiplicity of our newly discovered telematic being. To put some semblance of adherence into this spiral of recombinant phantasms, a new "pataphysics" of identity and social relations may be the appropriate strategy.

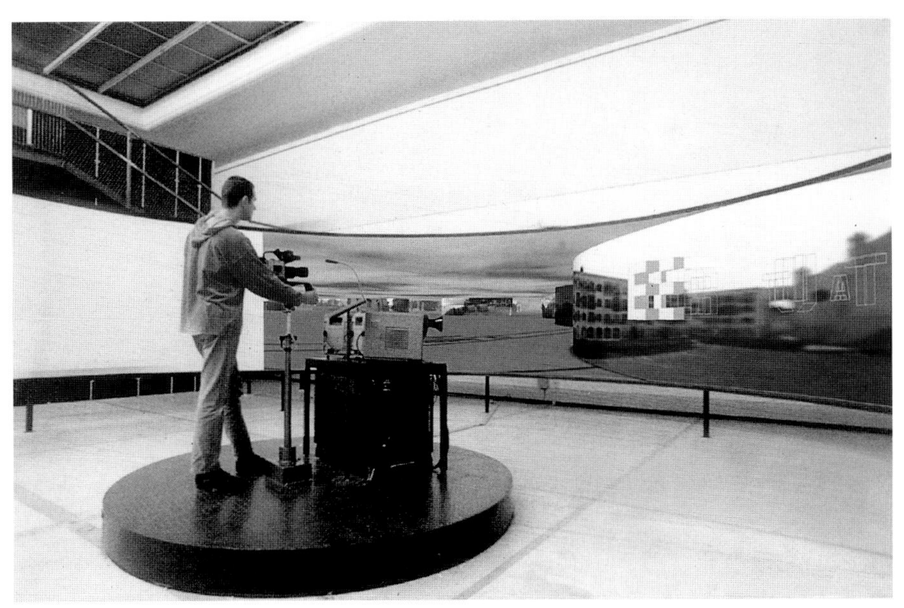

Fig. 3. Place – a User's Manual. © 1995 Jeffrey Shaw, Collection of the Stedelijk Museum Amsterdam

As we project and propel ourselves into these media machines, we must increase their complexity at a furious rate to try to make them embody the range of our human desires. But these machines are not us, and as their complexities increase to the point where they begin to take on idiosyncratic forms of their own, we lose control and become fascinated observers of their capabilities and anomalies. As artists we shift from the position of creator to that of critical cultivator, first searching to comprehend the possible meanings that emerge from this accumulation of nano circuitry and indeterminate layers of code, then trying to reconstitute those emergent phenomena in such a way that they can become part of an evolving cultural discourse.

What remains are questions concerning actual demonstrable confluences of art and science. There is evidence that contemporary scientific developments have influenced certain currents in art – Impressionism, Cubism, Constructivism and Futurism. There is less evidence that artistic works have directly influenced science. One could argue that the primary intentions of these two forms of creative endeavour are fundamentally different – the one aimed at the object and the other at the subject. In my opinion the question is more relevant at the level of industrialization of scientific knowledge – in the social arena where science becomes transformed by the specific agendas of exploitative concerns. All artists working with the new technologies have struggled with the rigid constraints imposed by hardware and software morphologies that have been configured by military, industrial and/or commercial attitudes. Here the artist has a most provocative role with respect to science and technology. That is to expose and even undermine those attitudes by creating a sensibility that posits and implements new relationships between what Siegfried Zielinski calls "media machines" and "media people". Relationships that embody the exploration and articulation of human complexities and experiences which is the essential domain of artistic research and practice. The challenge is to apply these artistic skills in the heterogeneous territory of the media machines.

The Unreasonable Effectiveness of the Methodological Convergence of Art and Science

Peter Weibel

Scientia Sine Arte Nihil Est
Ars Sine Scientia Nihil Est
Jean Vignot, 1392

I. Paragon

If we speak about art and about science we have first to define what kind of art and science we mean when we speak of them. Do we speak of the social sciences or the human sciences or the natural sciences? Do we mean the art of images or of sound or of space?

Since the beginning there exists the famous paragon dispute in the arts, the question what form of art is the guide line for the other arts, which activity of art can claim a primacy over the other art activities. There have been different answers. For centuries architecture claimed to be the mother of all arts. During the romantic movement music was the model for all the other arts. In the visual arts it is painting, which demands to be superior to photography and film as an art form.

This paragon dispute is not unknown to the historian of science. Here we have the first parallelism between art and science. The human sciences or the social sciences are not the model science any longer since many decades. Since the age of the Industrial Revolution the natural sciences have taken over as model science. But like in the visual arts again the question remains who can demand priority – physics, mathematics, chemistry, biology.

So when we speak of the convergence of art and science do we mean that of mathematics and music or physics and painting or biology and sculpture or linguistics and architecture? Is it actually reasonable to compare these different disciplines? Can the convergence of art and science even be defined on the level of similar disciplines? Naturally we do have some practical examples available to affirm the last question.

We could use elements of biology like symmetry, spirals, anti-symmetry etc. to explain the formation of patterns in the visual arts. We could show that for centuries biology served as an external morphology for the visual arts. In the time of the mimetic arts artists depicted the biological world and in the time of abstract art artists used the patterns of organic forms as internal principles for the organization of the pictorial elements.

We could show how a part of biology and a part of physics – the structure of the ear and of the sound waves – could be used to legitimate the laws of harmony in the art of sound. Above all geometry and mathematics have since the Renaissance served as central players and role models for the evolution of the visual arts and architecture, from the golden section over the laws of proportion to the contemporary fractal geometry[1]. The experimental psychology of perception[2] developed in the 19th century and the gestalt theory[3] of the beginning 20th century had an enormous influence on the arts, from impressionism to op art, from the foundation of cinema to photography. In many cases art in this context only serves as illustration of the laws of perception, from chromatic analysis (Seurat)[4] to stereo-cineticism (Duchamp)[5], so that we duly can call in question its legitimation as art. To compare Marey's scientific photography[6] of moving objects with the paintings of the futurists, Mach Bands[7] with paintings of Mark Rothko, Kanizsa's subjective contour patterns[8] with paintings of Vasarely, Attneave's stimuli[9] used in his studies of perceptual orientations with line drawings of LeWitt etc. etc. delivers triumphant proofs not only of the convergence of art and science but also of the primacy of science. In respect to the scientific investigations, analysis and discoveries many of the art works look trivially, delayed, derivative, just a popularization of scientific thoughts.

But just as these art examples likewise this method of comparison could look trivially. On a methodologically higher level the artists themselves compared the convergence of art and science and analyzed how science has influenced art. Especially from physics and its by-product technology art has been influenced deeply in the 20th century as the artists themselves made evident. Around 1927 Malevich had put together charts where he paralleled the development of the technical civilization and the development of the visual arts. In one of these charts, investigating the relation between the painterly perception and the environment of the artist, he compared cubism with the technology of factories, futurism with the wheel technology and suprematism with aviation[10]. Malevich was very well aware that every trend in art is linked with the view of the world, the totality of a particular culture. In his magazine "L'Ésprit nouveau" (1920–1925) and in his books of the same years "Ver une architecture", "Urbanism" and "Almanach de l'architecture moderne" Le Corbusier showed constantly cars, airplanes, turbine engines and ventilators etc. in parallel to buildings. The almanachs of architectures looked like the almanachs of machines. And the books of the avant-garde art looked also like books of machines, e.g. "Das Buch neuer Künstler" (Book of New Artists) by Ludwig Kassak and Laszlo Moholy-Nagy (Vienna 1922) where paintings and sculptures were confronted with images of machines. These publications anticipated and influenced the Bauhaus-books, testimonies of art in the industrial age, testimonies of art under the influence of technology. Especially in the 50s and 60s the influence of technology and philosophies of technology like cybernetics and system theory boomed and led to electronic music and electronic imaging. A new branch appeared on the tree of art: media art. This expansion of the field of the visual, starting with the painter Vermeer, who used a camera obscura to make his paintings of Delfft, was strikingly enhanced by the advent of photography, film, video, and computer. Machine supported generation and distribution of images, dislocated the place of the visual from easel-painting to screen. Typical for the convergence of art and technology in the 60s which led to an immense expansion of art practices and concepts and even to a theory of expanded arts was a movement called "Experiments in Art and Technology" (EAT), where artists like Robert Rauschenberg collaborated with engineers like Billy Klüver.

But not only the natural sciences have influenced the evolution of art but also philosophy and linguistics, especially the philosophy of language, psychoanalysis, semiotics, structuralism, and post-structuralism. From surrealism to concept art, from the bricolage tendencies (the artist as anthropologist, as field-worker) to the British group Art & Language (from the 60s and 70s) as prototype there is plenty of evidence of this influence[11].

Nobody therefore will dispute that science – be it sociology, biology, mathematics, medicine – indeed influences the arts. The question how much art is influencing science, which would be the basis of a true convergence, is a question much more difficult to answer and mostly negated. Is it correct to say a true convergence must be mutual? Is convergence the same or something different as mutual influence? And what would be the consequence of the confirmation that art is influenced by science? We will try to answer these questions.

II. Influences of Science unto Art

First we would say that the thesis that confirms the influence of science unto art is based on the difference of art and science, otherwise we would not speak of influence. So this thesis which apparently tries to melt science and art in fact is constituting the difference between art and science. Therefore the long chain of partial influences between art and science since the Renaissance which I have sketched paradoxically just confirms the difference between art and science. Our investigation on the convergence of art and science cannot have as product the difference between art and science. Therefore our analysis of the relationship between art and science must go beyond the investigation of examples of influences. The prior main argument how science is influencing art, the examples, is avoided by us as irrelevant and dangerous, because it is not only increasing the difference between art and science, but this argument is also a major danger to pauperize art as being under science. The question how much art and science are approaching each other must therefore be answered on the level of methodology. Are art and science parallel universes[12] which communicate with each other, which converge, which are permeable, or are they completely separate worlds, how it was asserted by the famous thesis of C.P. Snow. But the arguments of C.P. Snow have been argumenta ad hominem. If we could imagine an individuum, intelligent enough and comprehensively educated, this individuum could move in both universes freely. Another argument against Snow would be there are not only two worlds, but n worlds, chemistry, mathematics, crystallography, physics etc. etc., which are separated on the level of individual capacities, not only the world of art and science, because it would be likewise difficult to find an individuum which is a professional expert in molecular biology, proof theory and physics, or an individuum which is at home in the arts and in the sciences. The universe of science is separated into many sub-universes very similar to the separation of art and science.

III. Art as Method

So our analysis, as we have said, can only be done on a methodological level. Therefore we have to compare science and art as methods. That means not only to accept for science to be a method (e.g. in the way of Descartes), but also for art to be a method. And there is already the first difficulty, because traditionally people do not think that art is a method. They like to believe that art is just the opposite. Art is praised not to be a method. Science is characterized by its methodological approach, art is believed to be unmethodological. Art be the country of absolute freedom, contingency, individual eccentricity etc. This is our first claim: art and science can only reasonably be compared, if we accept that both are methods. This does not mean that we declare that both have the same methods. We only want to declare that both have a methodological approach, even if their methods are or can be different.

Especially since the Renaissance art has legitimated itself as a method, as research, even as collective research. In dispute with architecture on the question of primacy, as a product of the paragon dispute, visual art was forced to find a scientific method for the construction of an image because architecture since Brunelleschi's discovery of perspective had found its scientific foundation and legitimation on a scientific method. Painting had to define itself as a scientific method not in rivalry to science but in competition with another art form, with architecture. Remaining unscientific painting would have lost the battle and confirmed the hegemony of architecture. To be taken seriously also the visual arts had to come up with a scientific method, with a methodological foundation. This method, this scientific construction of the image was the central perspective, the re-presentation of three-dimensional space on the two-dimensional plane of an image. Therefore the painters called the perspective: "la costruzione legitima". This legitimate method of construction, this scientific method, how to organize the elements of an image, was seemingly broken down in the 19th century by the impressionists and Paul Cézanne. But this assumption is based on an mistake. The dispension of perspective through Cézanne did not entail the dispension of a scientific method. The scientific method was just replaced by another one. The science of perspective was substituted by the science of colour. Therefore Seurat could interpret his art as science, as a kind of painting "with the means of the time and in concordance with the science of the time". Aesthetics as a doctrine of rules was the most important step in the history of art to create art as a method similar to scientific methods. Aesthetics, founded in 18th century, is nothing else as the foundation and legitimation of art as a method. Should we say as a scientific method? And if we say yes, a scientific method, how then could we make a difference between the scientific method of visual arts and the scientific methods of natural sciences? As the products of science and art are so different, should we therefore also speak of different methods? In the beginning, geometers have been painters and the methods of architecture, visual arts and geometry, based on perspective, looked very much similar and therefore similarly scientific. But in the course of the evolution, the methods of science and arts have diversified as well as in the arts and in the sciences and between the arts and the sciences. It happened that the difference between art and science became a gap, and it became nearly impossible to speak of art as science and of science as art. Religion, science and art became in the age of the Industrial Revolution completely different domains and (after Hegel) different steps in the evolution of mankind. So it seemed even unreasonable to speak of art as a method, not to say, as a scientific method. Art and science looked like enemies, even when there was still a secret link. Many artists, starting with their opposition to the Industrial Revolution, also condemned science and technology. So art was established as a counter-science. Romantic art was a call for the irrational and against the rational of science and technology, with after-effects even up to Joseph Beuys.

IV. The New Nature of the Visual

Traditional art after 1945 obscured the relationship between art and science, only the media arts kept the dialogue and the contact with the sciences, because they themselves were based on technology. Under the influence of this new methodological approach between art and science based on technology in the 60s and 70s also the progressive visual arts innovated themselves in relation to the media. In the 60s painting started again to define itself as a method by defining it to the methods media, in relation which have constituted themselves as a methodology. The role which was once played by architecture to force painting into a scientific methodological foundation was now played by the media. "La costruzione legitima" was not perspective any more, even when the media arts re-introduced problems of perspective on a new level, but the methods of the media itself. The way the media have changed the material and the place of visuality was something, which could not remain unnoticed by painters who wanted progression. Therefore progressive painting founded its methodologies on the methods of image-construction of the media. So the media and media-related painting had a profound influence on our Western conception of the image.

The primary place of the visual is not any longer the image. This paradox wants to demonstrate that the horizon of the visual is greater than that of the image, provided that we understand as image the classical easel-painting. One of the main consequences for the concept of the visual has been that it was able to detach itself from its historic location, the tableau. Since the invention of photography in 1840, the image is no longer dependent on the canvas and oil paints, or a specific location; rather its location is flexible, it can shift, be transposed, e.g. onto the screen. The visual has become something more universal than the image. The classical image has become a passage of the visual. The visual is wandering through different media. The visual thus has successfully taken possession of new contexts; it has taken over new material, technical, urbane, cognitive contexts. It is precisely this new kind of interaction with the media, which holds the most fascinating promise and heralds a very open future for the image.

With the separation of the code from the carrier of the message, of the idea behind the picture from the carrier medium, which first became possible when telegraphy was invented, images no longer rely on tableaus for their transmission, but became free to exploit all media as carrier medium. In the 90s, this separation of the message from the message carrier, of the carrier medium from the code, has also begun to permeate and diffuse the art of painting. Our concept of the image has shifted; originally associated with the tableau and then with the photograph, it is now associated with TV and computer screens. With this altering association of the image with different carrier media our conceptualization of the image itself has transformed.

The classical conceptualization of painting is based on the assumption that there is a direct relationship between paint/colour and the canvas, with artistic subjectivity as the only mediator; artists are believed to access the canvas directly, just as the constituents of painting are assumed to be immediately present, much like in Henri Bergson's notion in "Essai sur les données immédiates de la conscience", 1889. In contrast with this view of painting as something immediately present, something that has no roots and no past – a view which dominated art in the first half of this century right up to the 50s and its informel movement – the 60s witnessed the evolution of a new concept which saw painting as a mediated art (e.g. Pop Art etc.) and postulated that art was derived from a source, i.e.., from the images of the mass media and art history. In the 90s, we see renewed efforts being made to link up with these earlier achievements and at the same time, to take the horizon of representation itself under closer scrutiny through critical experimentation. Painters do not simply react to the images created by the media and the history of art these transmit (e.g. in the printed media), but they go further by trying to anticipate the impact of these images on art, and to come to terms with it in, and through, their paintings. Instead of painting from nature as their 19th-century counterparts did, their images are derived from a "mediatized" world, i.e.. a world mediated by the mass media. Their paintings reflect a world, which has been transformed by technology and is dominated by the abstract codes and sophisticated effects of the mass media.

The location of the visual is the context, the historic and cultural code, the existing visual world. The visual is created in the war of images, the flood of images, in the world of the media. The visual is the result of multiple mediatized processes. Mediatization can exploit the technical media, but it can also refer to the codes derived from the history of painting. The visual today is defined as mediated visuality or visuality in context.

V. Art and Science as Social Construction

We can conclude from all our excursions that the methods of science and art can be on the one side very similar, even by artists who claim to be opponents of science. The installations of Joseph Beuys for example can look very close to a museum of natural history. Some installations of other artists can look very close to the installations in an ethnological museum. A work by a conceptual artist can nearly be identical with the work of a logician or a linguist. Even the work of a seemingly subjectivist eccentric like Bruce Naumann has its methodological basis, for example mathematics. Naumann has studied between 1960 and 1964 mathematics, physics and art. For two years he has studied the famous work of Kurt Gödel on formally undecidable sentences by writing it down on paper word for word. Still today (1980) in an interview he says: "I had a curious faible for mathematics. Still today I feel myself in my work drawn – how should I call it – to the abstract or intellectual. Anyway, what interested me at mathematics was more the structure than the concrete problem-solving, because this has more to do with philosophy as with practical questions."[13].

Artists are attracted to the methods of science, because they feel the structural similarity to the methods of art. The methods of art are different than the methods of science, but still methods. On the basis of the different methodologies and their parallels and differences art and science should be compared. On this level – which is higher than mutual influences – one of our first questions can be answered. Art and science are convergent on the level of methodology. When we speak of convergence between art and science, we think of methods. This methodological convergence is even mutual. Not on the level of product science is influenced by art, not on the level on references, but on the level of methods. Because any time when science develops the tendency that its methods become too authoritarian, become too dogmatic, science turns to art and to the methodology of art which is plurality of methods. Methods of science are characterized by doctrines, by enforced methodology. Art lives from a tolerance of methods, of a diversity of methods. Freedom of art means also freedom of methods. In his book from 1984 "Science as Art", Paul Feyerabend discovered the analogy between the plurality of styles, described by the Austrian art historian Alois Riegl, and the plurality of methods he wanted in science. In his work "Late Art Industry from Rome" (1901) Riedl developed his theory that there is no progress and no decay in art, only different forms of styles. This styles have their own laws and cannot be compared with other styles. Has the Renaissance developed linear perspective as the only legitimate construction of an image does this not say, that other ways of constructing an image are illegitimate. Each new specific art activity creates its own laws and styles, independent from other laws and styles. Alois Riegl's aesthetic relativism corresponds with Feyerabend's epistemological relativism. His famous "anything goes", the plurality and diversity of methods in the arts, should also be possible in science. Feyerabend went even so far, to say, that science should behave like the traditional image of art, to be without method. In his book from 1983 with the characteristic title "Against Method" he voted for a science for which any method is valid (this is the meaning of "anything goes"). In his book "Science in a free Society" (1978) he voted for a methodologically pluralistic, democratic science, which in its degree of freedom can be compared to the method of art. This is the meaning of his sentence "science as art": science as free method. We see on the level of methods science sometimes can be compared to art and science converges with art. This is a tradition, by the way, which goes back to Ernst Mach and Ludwig Boltzmann. In his famous essay "On the Development of Methods of Theoretical Physics" from 1899, Boltzmann considered the possibility that two theories, totally different, can be equally valid. The confirmation that a theory is the only correct one, is only a subjective impression. Plurality of theories and models are known in the history of science. When science becomes in its development too much a totalitarian doctrin, a monopoly of discourse, than it has to become an art, to turn to art, to reassure itself as a plurality and diversity of modes and methods. Feyerabend tried to show the mechanisms of the social construction of science which are comparable to the mechanisms of the social construction of art. A community of institutions and individuals (artists, critics, curators, collectors, galleries, museums) creates a social consensus what art is. Likewise a community of institutions and individuals agrees consensually what science is. From time to time there are individuals who challenge the consensus and propose a change of paradigms. In his books "Laboratory Life" (1979) and "Science in Action" (1987) Bruno Latour shows that our idea of modernity is based on a strict distinction between natural and social instances. But he showed that the distinction between culture and nature, between society and natural sciences is not totally clear. How much social instances helped to construct nature and how much have the natural sciences and their ideas of nature constructed culture and society? He claims that in reality there was an exchange between society and nature and art and natural sciences, which has created hybrids. The transfer of social categories on the construction of nature through modern natural sciences has also transformed our society. The transfer of natural categories on the construction of culture through modern society has transformed and defined our ideas of society and culture. There is a mutual transfer going on between society and culture, nature and natural sciences, between culture and natural sciences. There is no objective nature any more, separated from social construction, and there is no absolute art any more, separated from social construction. Art and science meet and converge in the method of social construction. Art as social construction and science as social construction converge in the postmodern field.[14]

VI. The Transformation of Art in the Technical Age

The transformation of art under the Industrial Revolution led not only to a machine-based art, but also to the machine-based generation of images, and machine-supported vision. The primacy of the eye, anticipated by Odilion Redon (1882), as the dominant sense organ of the twentieth century is the consequence of a technical revolution that put an enormous apparatus to the service of vision. The rise of the eye is rooted in the fact that all of its aspects (creation, transmission, reception) were supported by analog and digital machines. The triumph of the visual in the twentieth century is the triumph of a techno-vision.

This can best be demonstrated by the interpretation of the word "video". The Latin word "video", meaning "I see", referred to the activity of a subject. Today it is the name of a machine system of vision. This turn shows clearly that we have entered a new era of vision, the technical vision, the machine-based vision. Machines generate, transmit, receive, and interpret images. Machines observe for us, see for us. The eye triumphs only with help of machines. This mechanical perception has changed both the world and the human perception of the world. With machine vision man has lost another anthropomorphic monopoly.

From the iconographical point of view, two formative events occurred in the nineteenth century. First, through the advent of the technical image the fundamental idea of the image was changed. Hitherto, "the image" was a painting. But with photography the image escaped into other host media. Visual culture was no longer limited to the study of paintings, but extended to the study of photography, film, and so on. Image and vision dichotomized. The result of this encounter between image and technical media was the birth of the visual.

Secondly, another division occurred, namely the separation of the body and the message, through the invention of telegraphy (around 1840). Prior to this each message needed a physical, material carrier (a horse, a soldier, a pigeon, a ship). Suddenly a message could be sent without material carrier. Strings of signs could travel without a body. The scanning principle (invented around 1840) turning the spatial, two-dimensional form of the image into temporal form is central here. The immaterial world of signs established the basis of telematic culture.

VII. Post-Ontological Art: Virtuality, Variability, Viability

We know that the common link between both the technological/visual media of film and photography and the art media of painting and sculpture lies in the way visual information is stored. These material carriers make it extremely difficult to manipulate that information. Once recorded, visual information is irreversible. The individual image is unmoving, frozen, static. Any movement is, at best, illusion. The digital image represents the exact opposite. Here each component of the image is variable and adaptable. Not only can the image be controlled and manipulated in its entirety, but far more significantly, locally at each individual point. In the digital media all the parameters of information are instantly variable. Once a photograph, film, or video has been transferred into digital media its variability is dramatically improved. In the computer, information is not stored in enclosed systems, rather it is instantly retrievable and thus freely variable. Through this instant variability, the digital image is ideally suited for the creation of virtual environments and Interactive Computer Installations. Here, the character of the image changes radically. For the first time in history the image is a dynamic system.

Dependency on the observer will be enhanced in a system where information is saved dynamically. The image is transferred into a dynamic field of instantly variable points controlled directly by the observer. The context according to which indeterminate variables will assume their formal shape is now controlled directly by the observer, composing specific images from a field of variables, a variable sequence of binary components. The event experienced by the observer will depend on machine-generated variables determining their apparent shape or sound. The digital signal is defined by its original neutrality. Subsequently, it is transformed by input at the interface, its technological context, into image or sound signals, into a specific event. The image is now constituted by a series of events, sounds, and images made up of separate specific local events generated from within a dynamic system.

When defining the image we must now talk in terms of sequences of events

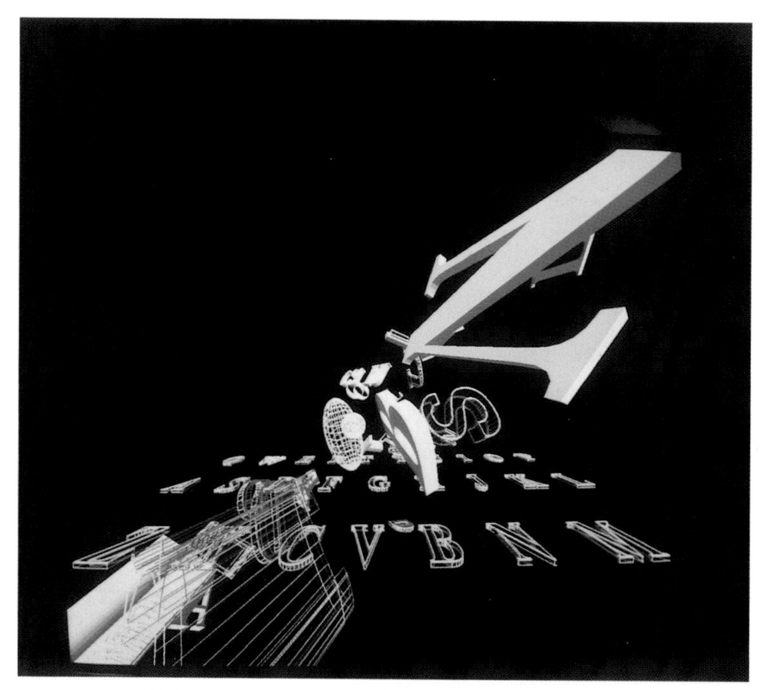

Fig. 1. © P. Weibel (1992) Zur Rechtfertigung der hypothetischen Natur der Kunst und der Nicht-Identität in der Objektwelt. Galerie Tanja Grunert, Köln

of acoustic and visual and visual variability and virtual information: of dynamic sequences of local (acoustic, visual, or olfactory) events. This vision challenges accepted formal aesthetic assumptions. The experience of events replacing the two-dimensional static image urges a radical revision of visual precepts, and the redefinition of context. The convention of a window onto a small part of a fixed event is becoming one of a door leading into a world of sequenced, multi-sensorial events, consisting of temporally and spatially dynamic experiential constructions that the observer is free to enter or leave at will. The quantifiable variables are now changed by their context. The context may constitute a different visual system, sound sequence, machine, human observer, distance, or pressure. We are able to construct ever-more sophisticated contextualities with the development of increasingly sophisticated state-of-the-art interface technologies (the human brain, limbs, light, movement, breathing may all transmit impulses via the interface toward the generation of context).

Though it is difficult to pinpoint the dominant influences on our perceptions from among endophysics, micro-particle physics, chaos theory, quantum physics, genetic engineering, or the theory of complexity, it is obvious that we are above all ruled by developments in what is known as computational science. Computers represent the most universal device ever available, just as their combination with the information sciences has advanced the most complex possible conceptual approaches. The current state of development in computer technology represents the pinnacle of technological and scientific research and development, which has accompanied a history of thousands of years of human evolution. It should thus surprise no one if our current perception of the human mind is that of a parallel-processing network computer.

Much will depend on the necessary incorporation of the major questions posed by historic aesthetic and conceptual assumptions. Thus, for instance, Plato's parable of the cave acquires renewed relevance in an age of simulated virtual realities: the existence of mechanical modules in the brain has now become apparent, though their exact role is still very much in doubt; similarly the discovery of the Butterfly Effect in chaos theory has made possible the translation of the observer problem in quantum mechanics into macroscopic dimensions. Clearly any future development of "brain sciences" will have to be vitally concerned with the role of the brain and mind in any conceptions of subjectivity and objectivity. Specific brain functions are already explored through the application and study of visual worlds dealing with interactive functions between observer and artificial universes via a multi-sensorial interface.

With the support of technology, traditional notions of our visual and aesthetic conceptions have been radically altered. The image has mutated into a context-controlled event-world. Another aspect of the variable virtual image is caused by

Fig. 2. © see Fig. 1

the dynamic properties of its immanent system. As the system itself is just as variable it will behave like a living organism. It is able to react to the context-generated output accordingly. The possible interactive nature of the media arts is therefore constituted by the following three characteristic elements of the digital image: virtuality (the way the information is saved), variability (of the image's object), and viability (as displayed by the behavioral patterns of the image). If we define a living organism as a system characterized by its propensity to react relatively independently to any number of inputs, then it follows that a dynamic visual system of multi-sensorial variables will approximate a living organism and its behavioral patterns.

Ultimately, the object of these new scenarios consists of and depends on binary information: objects, states, experiences are recorded and saved on data carriers after their transformation into binary code. Thus, the new worlds are virtual worlds. Through the retrieval of such binary data by algorithmic means, the instant manipulation of their content has become possible, and the object has become variable. In any virtual world its state, as well as that of its represented objects, may change according to either intrinsic simulation algorithms, or the reaction to external observer-generated inputs. The term "viability" is applied by Radical Constructivism to complex dynamic systems that are able to change their state autonomously via feedback reaction, and can react context-sensitively to varying inputs from their surroundings. In this sense viability denotes the possession of lifelike properties with the development of lifelike behavior. The digital trinity of saved virtual information, variability of image-object, and viability of image behavior has in fact animated the image through the generation of a dynamic interactive visual system. In new-media art installations it is possible to incorporate one or several human observers into computer-generated virtual scenarios via computer-controlled junctions in the form of multi-sensorial interfaces. The traditionally passive role of the observer in art is thus abolished; he turns from a position external to the object to become part of his observed visual realm, whose virtual scenarios will react to his presence and will in turn effect a feedback from him. The interactive installation has undermined our traditional assumptions about the image as a static object.

Thus, instead of the conventional world of the picture we have a universe of "free variables" floating in specific event-worlds, which can be comprehensively filled or replaced, and which interact with one another. The image has turned into a model world, auto-catalytic as well as context controlled. The animated image constitutes the most radical challenge of our classical visual notions of image and representation.

References

1 Schmidt WM (1983) Mathematik als Kunst. Ulmer Universitätsreden
Morse M (1951) Mathematics and the arts. Yale Review

2 Helmholtz H von (1867) Handbuch der physiologischen Optik. Leipzig

3 Ehrenfels Ch von (1890) Über Gestaltqualitäten. Vierteljahresschrift für Wissenschaftliche Philosophie 14: 242–292

Bühler K (1913) Die Gestaltwahrnehmungen. Experimentelle Untersuchungen zur psychologischen und ästhetischen Analyse der Raum- und Zeitanschauung. Stuttgart
Wertheimer M (1912) Experimentelle Studien über das Sehen von Bewegung. In: Zeitschrift für Psychologie 61: 161–265
Wertheimer M (1922) Untersuchungen zur Lehre von der Gestalt. In: Psych. Forschung I
Fischer KR, Stadler F (Hrsg) (1997) "Wahrnehmung und Gegenstandswelt". Zum Lebenswerk von Egon Brunswik. Springer, Wien New York
Brunswik E (1935) Experimentelle Psychologie in Demonstrationen. Wien
4 Homer William I (1964) Seurat and the science of painting. MIT, Cambridge
Rood Ogden Nicholas (1879) Modern chromatics, with applications to art and industry. New York
5 Benussi V (1912) Stroboskopische Scheinbewegungen und geometrisch-optische Gestalttäuschungen. In: Arch. f. ges. Psychol. 24: 31–62
Musatti CL (1924) Sui fenomeni stereocinetici. In: Archivo Italiano di Psicologia 3
Musatti CL (1929) Sulla plasticita reale stereocinetica e cinematografica. In: Arch. Ital. Psicol. 7
6 Marey EJ (1887) Photography of moving objects and the study of animal movement by chronophotography. Scientific American Supplement, Feb. 5, 1887
7 Ratliff F (1965) Mach bands. Quantitative studies on neural networks in the retina. Holden Day, San Francisco
Ratliff F (1972) Contour and contrast. Proceedings of the American Philosophical Society 115: 150–163
8 Kanizsa G (1976) Subjective contours. In: Scientific American 234: 48–52
9 Attneave F (1959) Applications of information theory to psychology. Holt, New York
10 Boersma LS (1988/1989) On art, art analysis and art education: the theoretical charts of Kazimir Malevich. In: Kazimir Malevich. 1878–1935. Russian Museum, Leningrad – Tretiakoff Gallery, Moscouw – Stedelijk Museum, Amsterdam (exhibition catalogue), S. 206–223
11 Rorty RM (ed) (1967) The linguistic turn. Essays in philosophical method. The University of Chicago Press, Chicago-London, 1992
12 Shlain L (1991) Parallel universes in space, time and light
Vitz PC, Glimcher AB (1984) Modern art and modern science. The parallel analysis of vision. Praeger, New York
Herrnstein Smith B, Plotnitsky A (eds) Plotnitsky A (Ed.) Mathematics, science and postclassical theory. The South Atlantic Quarterly, Vol. 94, Nr. 2, Duke University Press, Durham
Hahn W, Weibel P (eds) (1996) Evolutionäre Symmetrietheorie. Selbstorganisation und dynamische Systeme. Edition Universitas, Stuttgart
13 de Angelus M (1997) Interview mit Bruce Naumann: In: Bruce Naumann. Image/Text 1966–1996. Kunstmuseum Wolfsburg, Cantz Verlag (exhibition catalogue), S. 121
Naumann B (1972) Work from 1965 to 1972. Los Angeles County Museum of Art, Los Angeles (exhibition catalogue)
14 Pickering A (ed) (1992) Science as practice and culture. The University of Chicago Press

Stafford BM (1994) Artful science. MIT Press
Feyerabend P (1984) Wissenschaft als Kunst. Suhrkamp, Frankfurt/M.
Knorr-Cetina K (1981) The manufacture of knowledge. An essay on the constructivist and contextual nature of science. Pergamon Press, Oxford

5. Chaos and Complex Systems

Reflections of Chaos in Music and Art

Gottfried Mayer-Kress

Abstract

Important, new insights in science typically have their reflections in other areas of civilization and culture in that they modify or even revise the general world-view or "Zeitgeist". We have witnessed more than two decades of rapid and turbulent developments of the theory of self-organization, synergetics, non-linear dynamical systems, chaos theory, fractals, and other scientific sub-disciplines attempting to describe the miraculous world of non-simple (complex, complicated, erratic, unpredictable) phenomena in our lives. This development has dramatically changed our way at which we look at the world. In this paper we try to illustrate some of these developments with the help of a few examples of interactions between chaos theory and post-modern music and art.

Introduction

When I started to study physics it was pretty clear where the frontiers of pure science were to be found: in the very large scales of cosmology or in the very small scales of elementary particle physics. The intriguing questions were to find out where the universe came from and what it is ultimately made of. Like many idealistic young scientists I decided to explore the smallest building bricks of matter, the sub-atomic particles. I soon realized that this scientific frontier is so far removed from any form of human experience that it does not matter anymore what words or language one uses to describe the objects in this remote world of elementary particles. High energy physicists followed the explorer tradition of giving landmarks in the newly charted territories names or random adjectives from the Old World of our everyday experiences: What would one call a theoretical object that appears to describe a fundamental constituent of matter but that nobody has ever seen and – according to the theory – nobody will ever see?

Quarks and Modern Science

Murray Gell-Man decided to name those objects "quark" although they have nothing in common with the fresh German cheese of the same name. The physicist borrowed the word from Joyce's "Finnegan's Wake" just because of the sound of it. It is interesting to follow the naming conventions that Gell-Man's colleagues have chosen for the different types of quarks that were used for the taxonomy of the zoo of subatomic particles: Initially it was just the bland "up" and "down" that quantum theorists have used to name the "spin" of atomic particles. It was still possible to think of this spin as something that describes a spinning top familiar from childhood. That was probably the last time that any analogy to familiar objects could be used. It soon was recognized that a mere distinction of "up" and "down" was not appropriate to categorize the different species of quarks. It still makes sense that the next quark was called "strange" perhaps with the hope that this object would remain an oddity. This hope was given up when the fourth quark became unavoidable, the "charmed" quark. At this point it was clear that one might as well use any term to name those outer-worldy objects. Some physicists named quarks number five and six "bottom" and "top", others preferred to talk about "beauty" and "truth". These six different properties are referred to as the different "flavors" of the quarks. Quarks also come in different "colors". Of course, neither "flavor" nor "color" are notions that have any intrinsic "meaning" in this subatomic world. In a way, high energy physics is the ultimate of modern, abstract science. There is also no natural way to associate sounds with the findings of high energy physics, and the visuals basically have not changed since the early days of the cyclotron: tracks of colliding and exploding particles or bell shaped curves that are fitted to notoriously small numbers of data points. By the time I worked on my diploma thesis the big science of elementary particle physics had become almost as predictable as industrial mass production: Hundreds of scientists and technicians worked on huge experimental installations that were collecting events. One could almost predict which group would collect enough events first to publish the resulting bell curve, name it as evidence for a new particle and stand in line for the next Nobel prize.

Modern Music and Art
Above I made the (physico-centric) claim that developments in science are reflected in other cultural domains such as art and music. Do we recognize the spirit mentioned above in modern music and art? From my limited exposure to both I would see very strong evidence for parallel developments: Both modern art and music have explored landscapes of sounds, colors, and forms that are so far from our naive experience of music and art as is the flavor of quarks removed from the flavor of fresh cheese. In the same way as elementary particle physics had worked out a clear strategy of how to make new discoveries and win Nobel prizes – namely to go to higher energies and build bigger detectors – modern composers and artist had a clear evolutionary fitness parameter: Compose sounds/paint in ways that is different from anything that has been tried before. In science we can take advantage of achievements of earlier researchers as long as we include a reference to their work in our reports. Composers and artists are less fortunate: They are not allowed to build on the work of Mozart and develop his compositions further. The composition would have a hard time to be recognized in the modern art/music community if it reminded the expert audience of any earlier piece of work. Being assigned to a "style" is the ultimate nightmare of a truly modern artist.

Chaos and Postmodern Science

The advent of non-linear dynamics and chaos theory suddenly changed the whole scene: One discovered basic science frontiers out there that are not found in the domain of the very large or very small but in the new dimension of non-linearity. We became aware that science did neither understand the movement of a few planets (the question of the future of our solar system was not known because of chaotic predictability limitations) nor the behavior of a child's swing or a dripping faucet. It was a type of "not understanding" that was fundamentally different from the lack of understanding of details that we used to arrogantly toss away from our fundamental science perspective. The slogan used to be that daily life (including chemistry and biology) is governed by electro-magnetic forces and we understood those to an incredibly high accuracy. We did not recognize that this form of reductionist understanding was basically useless in making any sensible statements about the behavior of real world systems that all share the property of being non-linear. In many ways this recognition caused a similar shock in our view of the world as Einstein's insight that it doesn't make sense to talk about time as an absolute entity.

It was a fascinating experience to look at last centurie's experiments with the eyes of non-linear dynamics and chaos. Experiments on non-linear oscillators that I did with my youngest brother in his high-school lab were repeated by grown-up scientists and published in a reputable physics journal. A sense of terra nova right in front of our nose spread not only in the science community.

Chaos, Computers, and Postmodern Art

Eve Laramee was successful in the art world with her work on natural shapes of corroding materials and oscillating quicksilver hearts. The connection between her work and the new chaos theory became evident in the use of fractal concepts that made it possible to talk about her objects of exploration. A common feature of many of those objects is their rough texture and branching forms, signatures of fractals. Modern science and their field theories could not contribute much in describing the complexity of the irregular forms of her objects.

In high energy physics the true heroes of the seventies used pencil and paper alone to come up with creative new solutions to the immensely complicated problems of axiomatic quantum field theories and quark confinement. Computers were considered lowly crutches for those who were not smart enough to do pure science. But if one wants to understand fractal objects even the smartest mathematicians don't get very far without the help of powerful computers. In 1983 I attended a conference – organized by Heinz-Otto Peitgen – where Jean-Pierre Eckman, an Swiss chaos scientist and mathematical physicist gave a blackboard talk on chaos and Julia sets. He tried to describe their fascinating, fractal properties with a thought experiment of coloring maps of three countries who share all of their border points: if you reach the border of, say, country Red you have the choice of either entering country Green or country Blue. And you have that same choice at every crossing at any point of the border. I was so fascinated by this description that I wanted to see what these objects actually look like. It took our mechanical ink plotter several hours to complete a drawing of the z3-1 Julia set but it was immediately clear that this object had an aesthetic attraction that cannot be compared to the scattered bell-curves of high energy physics experiments. The response of the art world to these new fractal objects was mixed: It received an enormous attention as some form of scientific pop art but the computer artists often are amateurs with a background in mathematics or science. When I showed color slides of Julia sets to an art audience the response was less than enthusiastic. Fractal objects created by chaotic systems were considered to be more decorative than authentic art. The main direct applications in art appear to be more in areas of book covers, posters, and postcards or in the vast area of realistically rendered artificial worlds in science fiction movies. In postmodern fine art the inspiration from chaos theory was more subtle. One could find symbolic representation of the new meaning of chaos in our world and equations of chaotic oscillators incorporated in postmodern paintings.

Visualization and Audification of Chaotic Systems

Chaotic dynamical systems not only have an interesting graphical representation in their manifestation as Julia sets or other fractal basin boundaries but also in many other projections of their structural or dynamical properties into a visual or auditory domain. The different options of representation fall into two basic categories: representation of structures in state space (corresponding to the phase space of classical mechanics) and in the parameter or control space. In the former the behavioral structure of one given (chaotic) dynamical system is represented, in the latter it is the organization of different types of dynamical systems in a space of parameters that controls the behavior of the system in the corresponding state space. The Julia sets mentioned above fall into the first category, the now well known Mandelbrot sets fall into the second category. Sets of the second category can also be seen as a map or atlas of sets of the first category: Every point in the Mandelbrot display will generate a complete Julia set. Whereas Julia sets describe the behavior of a dynamical system in a boundary situation, attractors constitute a state-space representation of a "typical" class of observed behavior. That means if one starts a computer simulation at a random initial point one has a finite chance to observe the chaotic attractor emerging from the time history of the system. (The chance of finding an initial condition that traces a Julia set is zero.)

Both of the above spaces can potentially be of a very high dimension with the corresponding problems of adequate representation. In a "Phase Space Ship" installation at the 1991 Ars Electronica festival we used both three-dimensional graphics and nine-dimensional sound to display the attractor of a chaotic Lorenz attractor that lives in nine dimensions.[1] In an interactive program developed by Gideon May we could project the nine-dimensional state-space down into a three-dimensional graphical space. The choice of the projection was influenced by the type of local time evolution that was observed acoustically within the nine-dimensional space. This was done with software that was developed by Greg Kramer and that mapped different mathematical dimensions onto dimensions of auditory perception. The following Table 1 illustrates this mapping together with the estimated degree of resolutions in each of the dimensions i.e. how many different values can we expect to discriminate by changing the corresponding auditory dimension. We know that in the visual representation we have a very high resolution, perhaps better than one tenth of a percent in two-dimensional images. We also know that our (relative) depth perception has a resolution that is significantly lower (probably less than one percent) whereas our best auditory resolution (pitch) is about 3%. Other auditory parameters such as detune or duration have an even more moderate resolution of about 25%. On the other hand we know that slight changes in auditory signals have a very large capability to catch our attention, say, in a monitoring task that lasts for a long time. For example if we drive in our car for several hours we can still detect slight changes in complex sound parameters that might alert us of some problems with the car's engine. The significance and opportunities of sonic representations of complex or chaotic systems is now widely recognized and discussed at specialized conventions such as the series of ICAD meetings that started in 1992[2].

Table 1. Sonic map used for the visualization/audification of a 9-dimensional chaotic system. Note the wide range of resolutions for each of the sonic dimensions. This is similar to the lower resolution that we have in the depth dimension in visual space

Parameter	Description	Range	Resolution
Speed	Speed of pulses	Slow – Fast	16
Envelope	Attack and Decay	Sharp – Dull attack	4
Duration	Percentage of time of pulse	Short – Long	4
Detune	Detuning of two pitch components	Unison – Out of Tune	4
Flange	Phase shifting of sound	Stable – Swirling	4
Pan	Left/Right stereo placement	Left – Right	8
Pitch	Fundamental frequency	Low – High	32
Volume	Amplitude	Soft – Loud	8
Brightness	Energy of upper harmonics	Dull – Bright	8

Sound of Chaos and Brain Waves

Since 1985 the non-linear dynamics and chaos community witnessed a (sometimes controversial) discussion about the role and interpretation of chaos for describing brain-waves (EEG,MEG).[3,4] In the area of music performance brain-waves have been used to control musical instruments or other sound generating equipment. The sporadic appearance and disappearance of regular oscillations can be seen/heard in many recordings of brain waves. It could even be shown that their complexity (measured as the fractal dimension of a reconstructed attractor) changes with mental states of the subjects for example while listening to musical sounds of different degrees of complexity. These sounds themselves were created using chaotic dynamical systems at different parameter values. In [5] we could demonstrate that there is a resonant response of the brain wave activity to the complexity of the musical signal. In agreement with the hypothesis the response of the brain was similar in the cases of very high (periodic signal) or very low (random numbers) predictability but it was distinctly different in the case of the intermediate predictability of chaotic systems. In terms of dimensional complexity the different response of the EEG to chaotic musical signals was expressed in a reduction of the observed dimension. This result is also consistent with the hypothesis that any cognitive or perceptive process in the brain is accompanied by a short time (~100ms) synchronous oscillation of neuronal cell-assemblies that participate in the activity. In [6] we describe an experiment that demonstrate this effect acoustically. The paradigm is that of musical instruments in an orchestra: If the musicians play their instruments independently most of the time but sporadically some of them agrees on a tune, then these localized (in time) synchronization events should be audible as harmonic events above an incoherent noise background. The results based on four EEG channels controlling four different MIDI instruments are not completely conclusive (the sound files are available on the www site of [6]) but very promising as a monitoring tool for high-dimensional structures in multi-channel EEG/MEG signals.

Chaos and Postmodern Music

There is a long history of using computers in modern music either to generate note sequences or for the synthesis of sounds themselves. In the special branch of aleatoric music, composers used random number generators to create unpredictable streams of sounds. From the perspective of non-linear dynamics and chaos it is clear that pseudo-random number generators are nothing else but special types of chaotic systems with their properties optimized to satisfy the criteria of "good" random numbers. That implies the absence of any observable regularity. In the context of chaotic dynamical systems good random number generators correspond to systems with large fractal dimension, large Kolmogorov entropy, flat spectrum and low predictability. In other words: If one listens to a sonic representation of those numbers they sound (to me) very boring. The absence of structure in the auditory display of random numbers can be detected by our ears very quickly and it leaves us with the recognition of static noise. It is therefore natural to ask what happens if we take "bad" pseudo-random number generators with lots of structure and display them with sounds. The best systems for this purpose are chaotic attractors since they incorporate both long term unpredictability and surprise as well as recurrent recognizable structures or themes. The fact that these structures never recur with perfect precision –this is one of the defining features of chaotic systems– can even enhance the aesthetic value of the display. The same feature can be found in improvisations of a given theme in classical music and in jazz. The type of chaos that displays this feature most prominently is known as "intermittency"[7]; its dynamical characteristic alternates between times in which we perceive a regular (periodic) pattern and irregular, chaotic bursts which occur at unpredictable time intervals, followed by another return to a slightly modified version of the theme.

The hypothesis that we perceive something as aesthetically interesting if we have a balanced mixture between recognition and surprise has, to our knowledge, first been systematically explored by the mathematician and dynamical systems theorist George David Birkhoff (1884–1944)[8]. Birkhoff defined the "aesthetic" measure or "feeling of value" as the ratio of order to complexity.

Sounds of Chaos

So far we have described the application of chaotic systems to musical structures that are on the level of note sequences. The sounds themselves were assumed to be generated or synthesized based on an independent method. Early chaos pioneers like Otto E. Rossler who integrated their non-linear equations on analog computers occasionally connected their signal output to the input of an audio amplifier and listened to the sounds that are generated by those systems. Typically one would hear a pure tone for a periodic attractor that would make a transition to noise for chaotic parameters. For the fine ear of a composer there are, however, many different nuances of sound timbre audible between those two extremes. If one could find a good interface to change the multiple parameters of a chaotic analog circuit, one would actually have a new instrument with a completely new repertoire of sounds. It turns out that the Chua circuit satisfies all these criteria in an optimal fashion: It has an enormous variety of different sounding attractors, is easy to build and its dynamics is controlled by the parameter setting of six electronic components. Since some combinations of these physical parameters produce identical sounds it is a challenging task to design a control interface smart enough that it allows predictable transitions between attractors of sounds of the desired quality. In Fig. 1 we plotted three characteristic audio parameters of the sounds as a function of two of the systems parameters that avoid the redundancy of the physical parameters. The fractal nature of the control surfaces becomes evident which indicates the difficulty to play such a chaotic instrument without the help of a (digital) control computer[9,10].

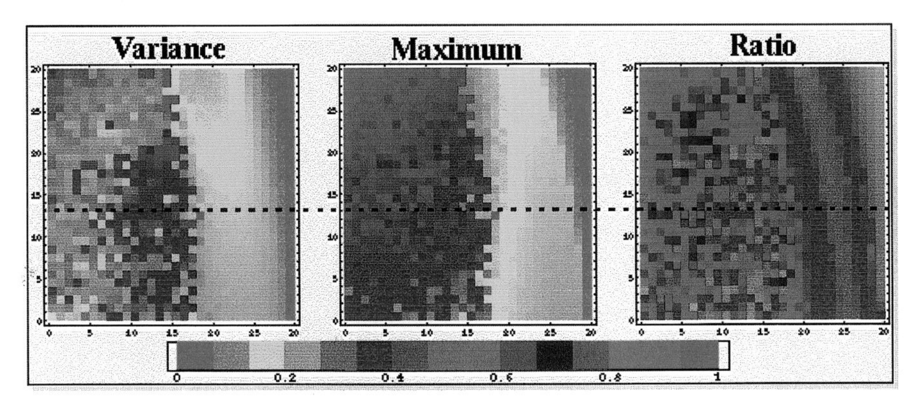

Fig. 1. Contour plot of variance (left), maximal amplitude (center), and ratio (right) of the signal from the Chua oscillator as a function of two of the bifurcation parameters. The ratio characterizes the content of intermittent excursions in the signal. © Gottfried Mayer-Kress

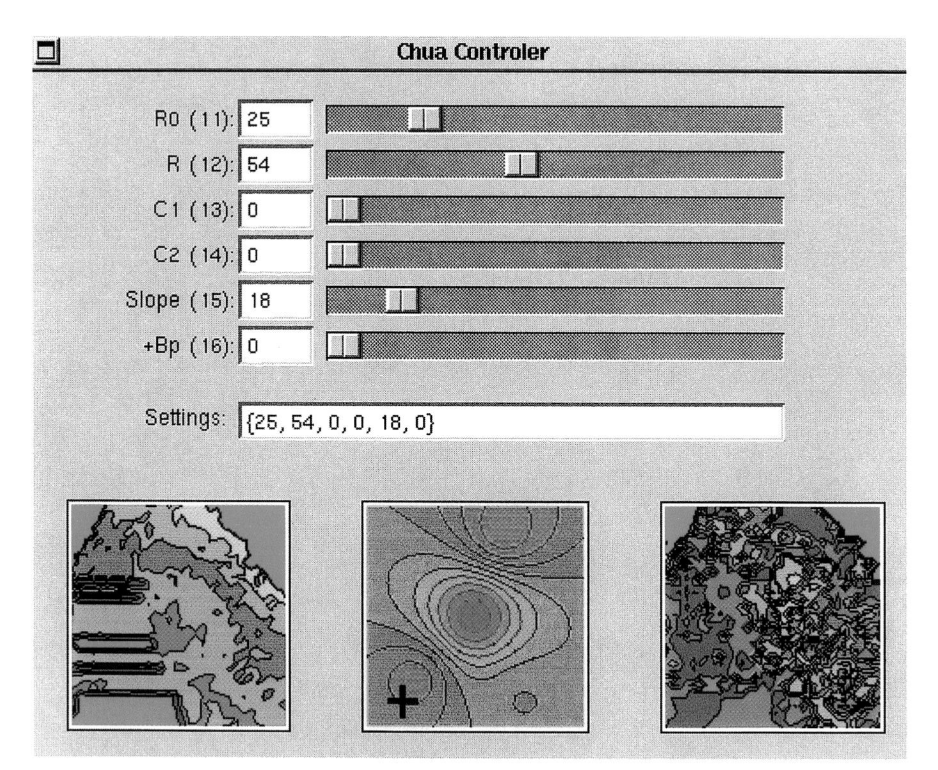

Fig. 2. Digital interface of the Directly Controlled Analog Chua Sound Module (DCACSS). © Gottfried Mayer-Kress

In Fig. 2. we see a simple example of a digital interface of the Directly Controlled Analog Chua Sound Module (DCACSS). The artist can control the analog Chua circuit either by moving one of the sliders that correspond directly to the physical parameters of the system or move a pointer across one of the two control surfaces at the bottom with color coding of acoustic outcomes from the circuit. More details and sound samples can be found in [9].

Future Outlook: Global Brains

The history of the past two decades have shown a dramatic change in the direction of scientific evolution back to the more direct experience of our daily life. This change of direction was influenced to a large extent by the enormous progress in both computer and communication technology. At the same time our daily life experiences have changed in a similarly dramatic way by exploring another new frontier: CyberSpace. There are several general rules in non-linear systems that might become applicable now that we directly experience the limits of our planet: Through the high degree of global connectivity that is becoming available for the first time in human history it now becomes likely that coherent structures or order parameters will emerge with shared properties and a coherence that has not been encountered before. This creates the possibility of a new form of global community including other intelligent species who share our ecological and environmental problems. We know that bounded systems under increased stress can exhibit transitions to chaos. But we also know that chaotic attractors can provide the opportunity to explore new domains in state space that can eventually lead to new and sustainable modes of behavior. A growing number of researchers is discussing the conditions for the transition to these globally coherent activities that bring also a new integration of the different domains of our civilizations and that is often summarized under the notion of a Global Brain.[11–13]

References

1 Mayer-Kress G (1991) Earthstation. In: Gerbel K (ed) Out of Control. Ars Electronica 1991, Landesverlag Linz, Linz, pp 147–156

2 Mayer-Kress G, Bargar R, Choi I (1994) Musical structures in data from chaotic attractors. In: Proceedings of the International Symposium on the Auditory Display (ICAD92), Santa Fe, NM Oct. 1992, Proceedings Volume XVIII Santa Fe Institute Series in the Sciences of Complexity, Addison Wesley, Reading (http://www.ccsr.uiuc.edu/People/gmk/Papers/ICAD92/ICAD92.html)

3 Layne SP, Mayer-Kress G, Holzfuss J (1986) Problems associated with dimensional analysis of electro-encephalogram data. In: Mayer-Kress G (ed) Dimensions and entropies in chaotic systems. Springer Series in Synergetics, Vol. 32, Springer, Berlin Heidelberg New York Tokyo, pp 246–256

4 Mayer-Kress G, Layne SP (1987) Dimensionality of the human electro-encephalogram. In: Koslow SH, Mandell AJ, Shlesinger MF (eds) Perspectives in biological dynamics and theoretical medicine. Annals of the New York Academy of Sciences, Vol. 504, New York, pp 62–86

5 Birbaumer N, Lutzenberger W, Rau H, Mayer-Kress G, Braun C (1996) Perception of music and dimensional complexity of brain activity. In: International Journal of Bifurcations and Chaos, 6(2): 267–278

6 Mayer-Kress G (1994) Sonification of multiple electrode human scalp electroencephalogram (EEG). In: Poster presentation and sound demo at: ICAD'94, Santa Fe Inst
http://www.ccsr.uiuc.edu/People/gmk/Projects/EEGSound/

7 Mayer-Kress G, Haken H: Intermittency in the logistic system. In: Physics Letters, 82A, 151–155, 981

8 Birkhoff GD (1933) Aesthetic measure. Harvard, University Press, Cambridge, MA

9 Mayer-Kress G, Choi I, Bargar R (1993) Sound synthesis and music composition using Chua's Oscillator. In: Proc, NOLTA93, Hawaii

10 Mayer-Kress G, Choi I, Weber N, Bargar R, Hubler A (1993) Musical signals from Chua's Circuit. In: IEEE Transactions on circuits and systems, Vol. 40, special issue on "Chaos in Nonlinear Electric Circuits", pp 688–695

11 Mayer-Kress G (1994) Emergence of global brains in cyber space. In: Knowbotic Research (ed) Nonlocated online: digital territories, incorporations and the matrix, Medien.Kunst.Passagen 3/94, Passagen, Koeln

12 Mayer-Kress G (1996) Messy futures and global brains. In: Kravtsov YA, Kadtke JB (eds) Predictability of complex dynamical systems, Springer, Berlin Heidelberg New York Tokyo, pp 209–232

13 Mayer-Kress G, Barczys C (1995) The global brain as an emergent structure from the worldwide computing network, and its implications for modelling. In: The Information Society, Vol 11 No 1: 1–28

Can a Program Force the Programmer to Reply

Otto E. Rossler

Abstract

It is a serious endeavor to look at the world as a machine.
The "web of causation" is much more daring a notion than is currently appreciated. The mind-set which gave rise to artificial worlds and the Internet is ultimately ethical and esthetic. It is argued that the 17-century nobleman who invented western enlightenment was enlightened also in the more artful eastern way.

The question mark has been omitted from the title because the answer is yes. The question is due to Descartes. So is the answer. Yet usually, this is not how Descartes is understood.

Descartes was poisoned at the age of 54, when he was visiting the Court of Queen Christine of Sweden (see [1] for the evidence). Queen Christine was one of the very few heads of state who abdicated while in power. After having just won the 30-year war over the catholic countries, she planned to convert to catholicism, and there probably were hopes that the country would follow her lead. In this undecided situation, Descartes' philosophical instructions could have exerted a stabilizing influence on the mind of the young (24-year-old) queen, a fact which no doubt endangered one or the other hoped-for outcome.

Thus, ordinary political scheming can be invoked as a potential reason for the murder.

At the same time though, Descartes' thinking[2] was unusual and destabilizing to society. As a nobleman from the country, he transported a Samurai-like feeling of independence and omnipotence to members of the lower classes. Moreover, he was almost unbearably sweet. If art consists of making enlightenment visible, he was an artist. And, living in Europe, he was killed for being an artist.

Other people had beautiful ideas at this time: Galileo, Kepler, Francis Bacon, Jacob Valentin Andreae, Calderon. His education was Ignatian. Ignatius of Loyola, the founder of the Jesuit order, was a gambler. He gambled for his life, with the creator, and taught this technique in his exercises.

Descartes later said that those who follow him will vanquish old age and eventually death itself (Hans Primas, personal communication). Only Everett, in modernity, nourished similar aspirations. His "relative-state formulation" of quantum mechanics (so the title of his 1957 paper[3]) opened up the way to a "world-change technology" as it may be called. If the world is an interface – a state relative to my own microscopically defined state –, then I can change the world by changing that state. The best way to do so is not immediately obvious – a Stelarc helmet with a magnet in it will probably be too simple. Since the only employer who dared to hire Everett was the Pentagon, all of his work from 1957 to 1983 (the year of his untimely death) is classified. He may become similarly important for the future as Descartes, whose writings were banned shortly after his death.

World-change technology nevertheless is still less than what Descartes aspired to. Such technique – the philosopher's stone – was only a pebble on the road for him. He strove for escape (without a Mustang). He had to, because a nightly dream forced him onto this path.

The dream was a fighting-dream in the spirit of Jacob. Jacob was a trickster with an irrepressible drive for greatness. He had snapped away the right of the first-born from his elder brother, Esau, and then from Isaac, his father. Now, Esau and his men were in hot pursuit to kill him and his family, and he had barely succeeded escaping with them to a precarious safety across the river Jabbok. In the evening, he glimpsed a desperate way of how he might be able to save his family – by awaiting Esau alone on the other side of the river. There he stood, fearing nothing any more, and – lo and behold – in the night there came a man to fight with him. It did not matter whether or not that man was Esau. The mightier he struggled before getting killed in the wrestle, the greater the impression would be on Esau – just as if he had fought him and his men with the same resolve. This spirit of compassion moved the god who had come to fight with him. After the night was over, Jacob had prevailed. As a reminder, he was given a lame leg and a new name (Man-Fights-God – Ishrael – or, perhaps, even Ishchael, Man-like-God, which has almost the same strokes in ancient Hebrew). Later, Jacob became famous also for a dream in which he hauled everyone to safety up the "Jacob's ladder."

This story was taken up in the nightly dream of young Descartes, in the night of November 10 to 11 of the year 1619, when he was 23. He was snowed in in Neuburg, a village near the town of Ulm in Bavaria. It was the first year of the 30-year war, and he was on his way back to the army in which he served as a mercenary officer. He stayed alone in a farm house, his bed and desk beside the huge central warm oven with enamel coating[4]. The dream was a lucid chain-dream, indistinguishable from waking reality, but terrifying. The wind played a major role in it. In the first part, the wind turned him around many times and he was left with a lame leg. In the second part, a book was blown in through the window onto the table beside his bed. It was the book of all possible knowledge. The chapter which drew his immediate attention was titled "Quod vitae sectabor iter" – what path of life am I to follow? Then there was a thunderbolt, the book was whirled away. After a while, the book was blown in again, but looked slightly different. When he feverishly opened it up, the decisive chapter had disappeared.

In the morning, he realized that not being able to tell a nightmare and reality apart, he would forever remain in the grip of the cruel dream-giving instance, a crazy man.

In this desperate situation, he glimpsed a straw to hang on to – the web of causation. Unlike the nightly reality, the waking reality might be consistent in its relational structure.

The world may differ from a hallucination in being consistent in its shadow part. What the ancient Greeks had valued the least – the "shadows" which alone survived in the underworld of the Hades, after colorful real life upon the Earth – could be made an anchor. The most fragile, most easily falsifiable hypothesis of history was born: the shadows might be of a machine-like consistency.

As long as this hypothesis was not yet falsified, he was safe, since the criterion allowed him to tell reality and dream apart. However, the unfalsified criterion was worth much more – it made him even. For if the world was a machine – and the others were machines, their brains were machines –, he was as privileged towards them as the dream-giving-instance that had almost killed him, was towards him.

Consistency implies exteriority – one's being totally outside like a programmer. Even though I did not program the others, I can look at them with a programmer's eye. I can also program analogs of theirs in a lower-level artificial universe, but most of all, they already are in a fish bowl for me – as long as the consistency hypothesis is still not falsified despite its ridiculous specificity.

Descartes did not know whether this logical dream of his would last ten minutes or ten days or a year. But he knew that as long as this hope was unfalsified, he could not rule out being in the role of a superprogrammer himself. And this gave him a chance – fairness: To renounce of misusing the infinite power of exteriority.

This twist made him enlightened. He re-invented compassion.

Jacob and Buddha had a Western companion.

The rest of the story is readily told. Science is a child of compassion. To check whether the world is consistent is a monk-like endeavor for many generations. It brings all kinds of fruit. The steel-fibers of analytic geometry (the mathematics Descartes gave the world as a present to check on the relational consistency) on the one hand enlarges the arsenals of the military to unprecedented proportions; on the other, medicine – the plumbing of the body as a deterministic machine – enables the survival in health of billions; third, the computer, with the vertical exteriority that it conveys relative to artificial universes, enables a new understanding of the world by analogy, and the horizontal exteriority of the Internet enables the first new evolution (and the first democracy). But most of all, the consistency check calms the mind by enforcing the image of infinite exteriority and superiority. As long as I cannot exclude being infinitely powerful I cannot refrain from acting fairly. Esthetics is inviolable. But: Is this not just the self-image of a southern-French minor aristocrat of the 17th century?

Emmanuel Levinas, the inventor of the technical notion of "exteriority," saw what Descartes saw: The face of the other is naked. It is talking. It says, don't kill me, don't leave me in my dying[5]. Only a self-styled demiurge can see the nakedness of the face. Compassion is made for gods.

Compassion is the ultimate assumingness. Buddha was a god.

Of course, he is only sweet Buddha – but no one is a greater challenge to the gods. If the programmers were not happy with the challenge, they would have intervened. Benjamin Franklin's test of the electricity coming out from the wet rope dangling from a kite sent flying into a thundercloud with his knuckle, was related in spirit. "No objection" is the ultimate reply in some cases. A program can force the programmer to reply – by not objecting. A mortal in a consistent world can execute infinite fairness even if it never existed before. He can refuse being an executioner. This is how Descartes' gentle soul functioned.

This book is called Art-at-Science. It could also be called Science-at-Art.

Acknowledgments
I thank Christa Sommerer for the kind invitation and Hans Diebner, Werner Pabst, Klaus-Peter Zauner, Walter Ratjen, Nils Roller, Rene Stettler, Detlev Linke, Adolf Muschg, Peter Weibel, Hidefumi Sawai, Keisuke Oki, Tom Ray and Siegfried Zielinski for discussions. I also thank the late Vilem Flusser for stimulation. For J.O.R.

References

1 Pies E (1992) Der (Mord-)Fall Descartes – eine kriminologisch-medizinische Untersuchung. (Descartes – a [murder] Case Story – a Criminological-Medical Investigation.) Brockmann, Köln
2 Descartes R (1641) Meditations on the first philosophy (in Latin). Soly, Paris
3 Everett H III (1957) Relative-state formulation! of quantum mechanics. Review of Modern Physics 29: pp. 454–462
4 Slightly condensed from the account given, from Descartes' lost manuscript "Olympica," by Baillet. Compare: Specht R (1980) Descartes, Bild-Monographie (Descartes, a Monograph with Pictures), Rowohlt, Reinbek; and Holz HH (1994) Descartes (in German). Campus-Verlag, Frankfurt
5 Levinas E (1987) Time and the other (Cohen R, transl.). Duquesne University Press, Pittsburg

6. Public Spaces

ICC (InterCommunication Center): The Matrix of Communication and Imagination

Toshiharu Itoh

Introduction

InterCommunication Center (ICC) is being developed as a prototypical information network oriented arts and sciences interface, a model 21st century museum.

Remarkable recent developments in electronic communications technology are laying the groundwork for a post-modern age successor to industrial present. We are still at a point however, where technology has made the more conspicuous progress, while explorations into the cultural and societal forms suited to an information-oriented world lag far behind. The ICC is being created to forward such dialogs between art and technology, especially in electronic communications. It is an attempt at an experimental depiction point for new visions of life in a post-industrial society.

Fig. 1. ICC museum entrance. © NTT-InterCommunication Center (ICC)

It follows that the ICC can not just exist as a museum where "objects" are stored and exhibited. It must be a software-based network point where "events" gather and interplay, giving birth to new information in the process. The ICC will function as a core node in a network of creative individuals at the leading edge of modern technology and global culture, with the ICC Matrix as its communications culture informational infrastructure. There will be workshops for information assembly and exchange, adding dimension to the scope of its activities. It is precisely this kind of activity that will allow the ICC's architectural facilities to show a continually expanding global resonance, remaining at the front of changes in the field, remaining vital to the community it serves, and remaining a leader in developing communications culture. The ICC will in itself provide a model of the structure and operation of museums in an information-oriented society.

For this reason, it was crucial that the ICC not wait for the completion of its architectural facilities to begin operations – that it begin operations through developing these software "events" as a network, which could develop momentum to reach its "critical mass". The ICC's architectural facilities are slated for completion in 1997. The ICC has already begun operations – in publishing, exhibitions, and other activities – to develop its own unique networks and databases, and have a full museum, up and running by 1997. But the Intercommunication Center will not begin from 1997. Rather, it began in 1991 – in a countdown of the last decade of the 20th century.

The Intercommunication Center Project Perspective

The rapid progress information-oriented societies has altered not only the face of industry and economy, but brought about great changes in the society and culture. Commemorating the 100th anniversary of its establishment, Nippon Telegraph and Telephone (NTT) created the Project InterCommunication Center to develop resources not only for industrial society, but to contribute to the higher levels of information culture. The ICC is a center for taking the leading edge electronic communication technologies core to NTT operations, and forming a broad-spectrum experimental link for how they will come to affect our society and culture.

The ICC is being built to become a first class cultural facility, competitive with any worldwide, as a core point for the people, places, and future that it involves.

Just as NTT has contributed thus far to society via communications technology, ICC aims to make a unique contribution to thematic communication – supporting creative endeavors and cultural development. We also look forward to the ICC offering exciting feedback into the world of technology.

The basic concept of ICC is:

"Communication", taken as the keynote issues of art, techno-science, and society – each "intercommunicating" real-time culture, creating a vision of 21st century cultural activity.

Key electronic processing technologies are racing forward, pulling with them the edges of our social fabric. The two major fields of computers and communications alone are exploding into and effecting changes in, our economic, industrial, and life styles.

Especially in the world of art, technology is playing an increasingly important role: testing the frontiers of creativity and aesthetic in providing new expressive forms, such as its capacity to produce images never before possible, and change the potentialities of the creator/viewer relationship.

The ICC is a cultural center based in the interchange of art and technology. Through new creative forms, the ICC hopes to awaken new forms of imagination relevant to our age – through new creations and experiments and communication perspectives which cross cultures and genres, transmitting and offering new cultural visions of the coming century.

Activities

The ICC will maintain an exhibition and event schedule based on its core theme of communication within techno-science and art, providing a variety of opportunities for reflection, observation, and experience. Developing important, leading-edge themes, in exhibitions curated by leaders in a wide variety of fields. Also, actively pursuing traveling exhibitions, and exhibitions developed in conjunction with other cultural facilities.

Fig. 2. ICC museum interior. © NTT-InterCommunication Center (ICC)

"Stage events" such as performance art, concerts, film and video screenings, including media events will form the core of a continually changing program of shorter term projects.

A number of on-going production activities and educational workshops will be held on the theme of art and techno-science.

Production support activities will include providing artists and engineers with studio space and equipment, related technological information, encouraging information exchanges with specialists from fields around the world in co-productions, and actively soliciting the contributions of the general public.

Educational activities will include classes for people of all ages, promoting intellectual and experiential knowledge opportunities to develop an understanding of the ICC's aims, and provide opportunity to the next generation of young artists.

With ICC's unique curriculum, starting with the one under development at the ICC New School, a number of symposiums and seminars will be held in an open forum for discussion.

Not limited to its architectural facilities, the ICC is developing into a variety of media to provide another stage for its activities. Beginning with the February '92 publication of "InterCommunication", the ICC's quarterly magazine, the ICC is active in a number of media including the production of software for television programs, videos, CD, LD, as well as continuing to develop its " Museum Inside the Telephone Network".

In anticipation of the opening of the ICC facilities, a number of projects are underway to lay a groundwork for understanding the ICC's image of art, techno-science and communications. These include networking activities, and knowledge-ware assembly, and a number of exhibitions and events, all building towards the completion of the ICC's architectural facilities in 1997.

For example the ICC constructed a virtual museum on the internet in 1995, it was called InterCommunication '95 "On the Web – The Museum inside the Network".

The preparation of a completely new and unexpected site for creativity has now been commenced by the fusion of the communications system based on telephone lines and the computer as both technology and media. This new site contains the potential to do away with geographical boundaries and cultural differences, as well as the momentum to transcend the limitations heretofore posed by material physicality. In other words, it possesses the potential to re-materialize and re-describe human beings.

When InterCommunication '91 "The Museum Inside The Telephone Network" was held, the word "Internet" was still nothing more than something from a distant world. But over the past few years, its "multimedia" environment has rapidly begun to permeate daily life. IC'91 used a network to create a museum, invisibly packaged, through telephones and faxes. With NTT InterCommunication '95, a visibly expanding and evolving museum came into being through the use of Netscape, Webscape and Hotjava on the Internet. Will the emergence of a museum of this form be capable of turning into a new system for preserving art and lengthening its lifespan? Or will it be able to transform art into completely new and previously unimaginable forms? These are the kinds of issues that will surely be questioned by this network. The technology of information communications is a "technology of consciousness" that belongs to the realm of the spirit and the senses more than to the realm of practicality and function.

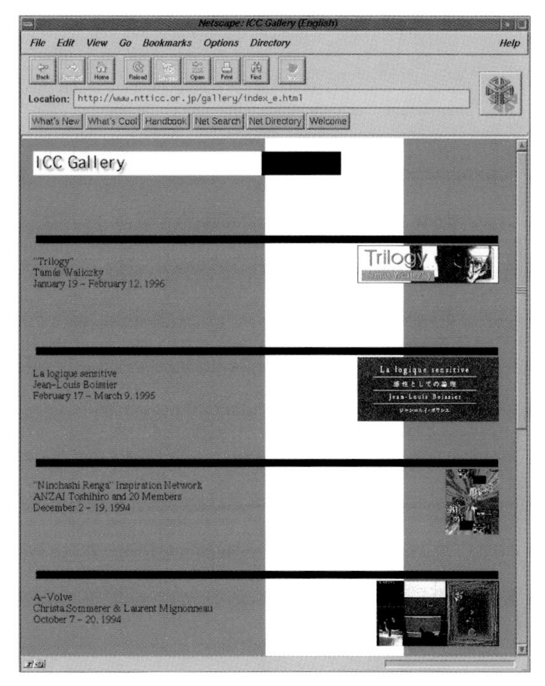

Fig. 3. ICC Gallery Web Page. © NTT-InterCommunication Center (ICC)

ICC Matrix

The ICC's "21st century software infrastructure" knowledge-(storage and development)-base, the ICC Matrix, will offer text, audio, and visual information on the history of art and techno-science in the 20th century, accessible by a unique ICC cross-vectored hypermedia system. Structured to be flexible to growth and change over time, the Matrix will not only be the ICC's core informational asset but also be fun to access. The Matrix will be accessible via electronic computer networks, able to exchange information, and remain accessible over networks.

The word museum takes its root from the temple of the Muses, the patron goddesses of the arts, which found its first realization in the Museon. The Museon was an educational institution for teaching the natural sciences, the arts, philosophy, literature and music, which made Alexandria, Egypt the cultural and literary center of the Hellenistic World. It included an art museum, vast libraries, scientific research facilities, botanical gardens, and even an astronomical observatory.

What was born there was an inter-cultural information space. Its establishment was closely bound to the dynamic movements of urban Alexandria, and its society. It was the new locus for collecting data, and creating information. It was a correspondence for transformation of the societies' senses of recognition and memory.

New concepts and forms of museum are always at the leading edge of the times, tied to transformations in the human intellect and changes in industrial organization.

The ICC was created from this perspective as an experimental grounds for such a "post-museum."

The information and knowledge assembled, curated and given form in the ICC are not gathered as 2D conceptual accompaniments to the artist and art work collections in other museums. Rather, they are in time to accomplish larger roles, to give forth the cultural values for the coming century.

As the ICC's "soft infrastructure" or knowledge base, the ICC Matrix is therefore bound to first examine the most activate magnetic fields – inspecting the cross points between the arts and sciences, as it ventures towards a new style of artistic database construction. The Matrix is to become the independent organic entity created of this information assembly and transference, as it continues to re-create itself in research and development, access developed with other institutions and individuals, its educational and network functions.

Employing an encyclopedic interactive hyper-media format was essential to achieve the kind of unique "soft infrastructure" which we had envisioned. André Malraux once described the museum as a location for the creation of new forms of knowledge and beauty. He created the phrase "le musée imaginaire" (the Imaginary Museum), to describe a dialectic which superseded issues of past and present, Western and Eastern aesthetic. What is not well known is that Malraux's prediction was that the successor to "le musée imaginaire" would be an "audio-visual museum." In short, the "le musée imaginaire" would be for prints, photographs, chiefly still, 2D works in reproduction, while the "audio-visual museum" would employ a variety of media technologies – not only to animate, re-dimensionalize, and render these originals in 3D – but to create new meanings and values from them.

This "audio-visual museum" would enable an approach which offered the creative and productive processes of the artists, and provided aspects of spontaneity and creativity lacking in the "le musée imaginaire". Malraux said that in the "audio-visual museum," not only the art works, and their production processes, but the vast, deep magnitude of phenomenon of the world that existed before they did would be exhibited, opening other paths, and issues of resource and cognition.

The concept of the "audio-visual museum" was something that Malraux became attached to in the later years of his life. It was never realized, and the phrase "le musée imaginaire" took a direction all of its own, becoming famous. However, in today's multimedia environment, we need to reassess this concept of the "audio-visual museum" from fresh new perspectives. Malraux's vision of an "audio-visual museum" was a return to a totality of vision, and communication, within multitudes of dialog and interaction. For example each art work, or each artist, or each countries art work... would find a new dimension of meaning in the meeting of these heterogeneous entitles. The ICC Matrix aims for exactly such an inter-zone a multivarious "audio-visual museum."

The ICC Matrix is the ICC's birthing body, a new type of body of audio, visual, and textual data on 20th century artistic and techno-scientific advancement.

Based in an encyclopedic collection of fundamental data, each individual bit

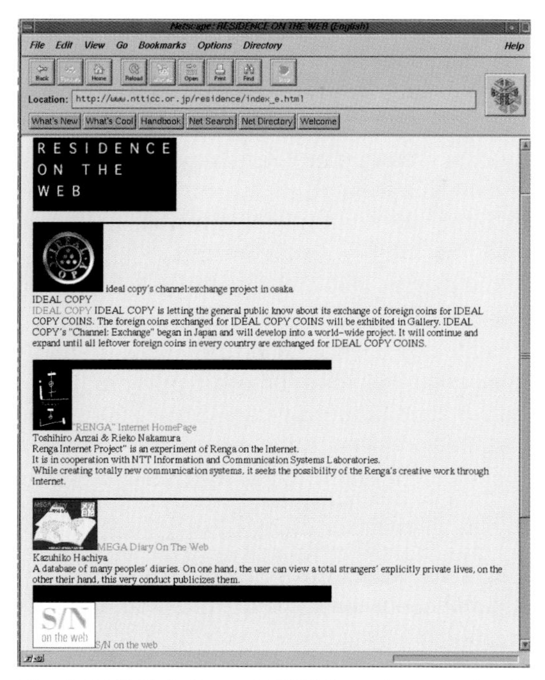

Fig. 4. ICC Web Page. © NTT-InterCommunication Center (ICC)

of data cross-linked so as to transcend the simple interpretations of genre, territory, or time period.

From the beginning of this time/space demarcation called the 20th century, colorful events and phenomenon have each risen independently. They scatter like water drops on a scalding steel plate, superficially separating, each appearing to quake, while, in fact they are merely flows given form by the same historical magma.

The problem with dividing art from science from music or from cinema is based in the fact that it is not as if each were an independent genre to begin with, but rather an intermingling of genres in complex flows, the creation of innumerable networks.

Therefore, what occurred in any period within any genre cannot be discussed without discussing it in relation to each of the other genres. Again, it is a mistake to consider the various events and phenomenon of the 20th century as though they occurred on a linear past / present / future absolutist time axis, or as though there was a cause first and then its effect, as part of a karmic law, for each multiidimensional co-habitant time could just as well be considered as intercolliding within space-time, and scattering. Perhaps we should call each of these the characteristics of this 20th century which we are living in. And it is about time that we free ourselves from the causal, analytical, reductive, linear time axis elemental explanations which have ruled our 20th century world view. The important thing is in finding clues from the words "change," and "relationship" which line up there genres, bridging the interior and exterior worlds, pilling the past on the present on the future, creating inroads between "real life" and imaginative power. Through doing so, we go beyond the time-space paradigms, realities, and concepts we have been so bound to and peer at the thrilling flows and dynamism that unfold in this universe known as the 20th century. The ICC Matrix takes just such a perspective on the 20th century arts and sciences exploring the relativities between data, images, and sounds through an accumulated information system informed by keywords. It will not be just another AV database, but a space for developing connectivities, based on the themes and concepts of the ICC.

To go further, this data and thesauri will increase continually, in continual transformation. This will be like the inside of our minds, where all manner of data are continually re-contextualized, or re-affirmed, while at the same time in perpetual dissection of those very contextualizations, in multilayered interrelations bringing about new associations, and forming new contexts and structures.

The visitor to the ICC Matrix will through entering this data accumulation, experience what did, and did not happen in the 20th century and gain clues as to what might be done next. The ICC Matrix will be an ever-expanding informational core for the 21st century, and through being in our possession we not only hold the data that we have assembled, but the seeds of unknown information to come. As a multilayered, accumulating, electronic memory system, it will be an inter-coupling, multiplying life form. The ICC Matrix will be a multidimensional storage of information related to the arts and sciences, whose oceans and forests we can freely navigate, and this, in fact, will be the basis for the "museum of the future."

New Art Needs New Art Venues – Art, Technology and the Museum: Ars sine scientia nihil est

Hans-Peter Schwarz

It was a French architect who told this doctrine to the Italian masons having serious problems with the construction of the cathedral of Milan in early 15th century Italy – the largest gothic building ever realized.

And the relationship between art, science and – later – technology in western art since then had a long and most of the time fertile history – in former times.

But when technology developed into industrialization, the romantic movement fought for the autonomy of art. And when industrialization became total and totalitarian – since the beginning of our century – philosophers, aesthetics, art-critics and – with certain but serious exceptions – even artists proclaimed an irreconcilable enmity between art and technology.

And since the middle of the century the relationship between art and technology was seen as a kind of "barbarous wedding".

And now, at the end of the age of industrialization and the beginning of the age of information, media-technology is supposed to be the enemy perhaps even the murder of art.

This rough description characterized till today the situation of the official art scene in Europe and with few differences in most of the other post-industrial countries.

Only very few museums or great art exhibitions like for example the last Biennale di Venezia are trying to integrate artworks deploying new media-technologies.

And even these rare approaches to media art are reduced on video-sculpture or video-installation, that means a type of media art, I wished to call "retroactive" media art, to distinguish it from her interactive cousin.

Retroactive media art like video-sculpture, video-installation or however these genres are called are still in the tradition of that playfully critical association with the obsolete remnants of technology to which the charm of the works of a Tinquely is indebted.

Or inherited the analytical gesture of reduction, the aesthetic game of nullification of Marcel Duchamp, who is often somewhat rashly claimed to be the grandfather of the current trends in media art.

Interactive media art, the most avant-garde position of contemporary art in general, realized the heritage of another ideology in modern thinking: the possibility, now shifted into the field of the technically feasible, the visitors of a museum or an art exhibition could react directly to the aesthetic structure of a work of art with the help of the computer controlled constellations of the machines, the interfaces, as it is now, seem to let the utopias of a whole school of thought in the modern age become a reality: the tendency towards the open work of art, or better, to "works of art in motion" as Umberto Eco put it.

And it is exactly this potential of interactive media art that will challenge the idea of the museum in a radical way.

The relation between each new electronic media and the traditional medium "museum" has never been smooth and this is still true today.

Characterized by the fear of being overcome by a flood of images devoid of any content on the one hand and on the other by dogmatically holding to the idea of possessing the only legitimate way of representing images, museums have almost completely ignored electronic image media for a long time.

This resistance is not so new. When we recollect how long it took for photography to be considered worthy of museums and when we recall the discussions about the use of audio-visual shows in exhibition, which are still controversial to this day, it becomes apparent that the basic orientation of the museum, which is in many respects conservative, is more likely to impede a forward-oriented acceptance of innovations.

However, the other side, the media artists and authors, is also scared of embracing this issue.

Especially the avant-garde in media art often view themselves as having directly inherited the legacy from their predecessors in classical modernism, which had declared the museum in its entirety to be obsolete and aimed to replace it by football stadiums or similar facilities.

Also, the idiosyncrasies of some productions especially in interactive media art with their structures designed more for an event of limited duration than for a permanent presentation, distort our vision of features, which the museum could indeed provide to the electronic avant-garde.

For this side too, there is still a great need to catch up in order to reach an understanding of the inherent conditions of each museum and artist.

But – whether we are fascinated or repulsed by this – the triumphal march of new media in society which can no longer be withstood seriously, has three entirely opposing or at least unyielding consequences to the museum: on the one hand, its subject area is expanded vastly by the host of media works of art, which are becoming more considerable. The second consequence of the new media for museums could be an expanding area of application. Even now numerous museums have installed their homepages in the world wide web. And once the much talked about information highways are finally open to traffic, art and culture trucks should also be available here.

But it is precisely this process of digitalization of museums which might be able on the one hand to break through the statics of museums, but on the other places extreme demands on consideration of the way museums view their role.

In general, the image of a museum is determined by walls derived from castle architecture, whether these be shaped historically or contemporary, by walls surrounding works of art or historic objects in order to protect them like treasure chambers which are to protect the most precious property of a nation, a region or even a private collector from the erosive powers of time.

However, this fortified image of a museum only dates from 200 years ago.

It developed when various bourgeois revolutionaries opened up the treasure chambers of their feudal adversaries and made them public property. It became more firmly established when in the course of the 19th century the bourgeoisie, because of a lack of a historical conception of their own, took on the artistic proofs of legitimization of the nobility re-engineering them for their own purposes.

In the course of the 20th century it survived all democratic and socialistic attacks almost undamaged.

Its real walls were even supplemented and reinforced by the ideological walls erected by educational privileges and cultural arrogance.

The fact, that the museum of the future could become a museum without walls, might even have to become one, is the result of two corresponding developments, an ideological one and a technological one.

Most of the attempts of the museum to open itself democratically towards society, which could be witnessed in recurring waves throughout this century, which is now coming to an end, have failed because they only enabled people to attend, but created hardly any emotional participation and almost never an intellectual one.

On the other hand, this field of experiments for a critical public, which the museum was in the past or at least claimed to be, now seems to gain its future form in the various discussion forums on the Internet.

Will the museum of the future cede its physical presence to the international data network?

Please permit me, conservative museologist that I am, to briefly consider what we have today: the museums of the present.

Despite all their serious shortcomings museums in their present form have one advantage vis-à-vis other mass media: they confront us with the alien, the unknown and also the embarrassing. Other mass media, above all TV and computer games, have an eye to political correctness and have turned the intimacy of private life into a public yard stick.

The museums, however, have resisted these attempts at legalization with surprising success. Maybe due to the personal obsessions of the people responsible for an exhibition or of a museum's curator, maybe due to the subjective passion of a collector or the long traditioned narrow-mindedness of a region or nation. Sometimes narrow-mindedness could be an advantage.

And even in those cases where works of art and historical-cultural artifacts are loaded with pedagogical claims because of cultural and educational policy, which threatens to take to their edge, objects in a museum – because of their authenticity – maintain enough of their aura to have subversive effects.

When the digital museum has already surpassed his foundations by far, I should like to ascertain one thing clearly here: a museum's existence on a network presents for me only one, albeit an extremely important and exciting expansion of the communication possibilities of the medium of the museum and not its dissolution.

A museum will have to remain that memorable place where we can reassure ourselves of our view of our own reality and the material reality of historical and artistic witnesses of history.

Maybe in the age of total simulations which leave us more and more uncertain in our experience of reality it is precisely the museum which could provide one of the last refuges for sensorial assessment of reality.

But this plea for the museum as it is now should only point out that the museum should be treated as an unique and idiosyncratic medium.

On the other hand it should be clear that the form of this medium has to be challenged completely. Our proposal for this new kind of museum is the institution we are preparing to establish in Karlsruhe south-west of Germany: The Center for Art and Media Technology (ZKM).

Fig. 1. East Facade of the ZKM. ONUK; © ZKM

I want to give only very few remarks on the organization. The institution combines the four important issues of art production and art reception through a close network of different departments: two research departments, one for music and acoustics, one for image media create a research and production environment for artists where they can work with the rapidly evolving media technologies. Its aims are to enable the practical research and development of appropriate media techniques and to promote the applications of these techniques in an artistic and social context.

A media library and a media theater are dedicated to the history and the presence of electroacoustic music, video performance and digital experiments in the performing arts in general.

In the school for art and design media theory and special forms of teaching should be developed and the museum of contemporary art presents the video arts together with contemporary works of painting and sculptures in the frame of the general development of art in our days.

And in the center of this complex organism is the media museum together with its virtual branch the Salon Digital. This part of the ZKM aims to focus all its activities and should allow a public access to these activities in the broadest sense.

The fundament of the concept of the museum is to give some answers to the question: "New Medias – Where are they from? –Where do they go to?"

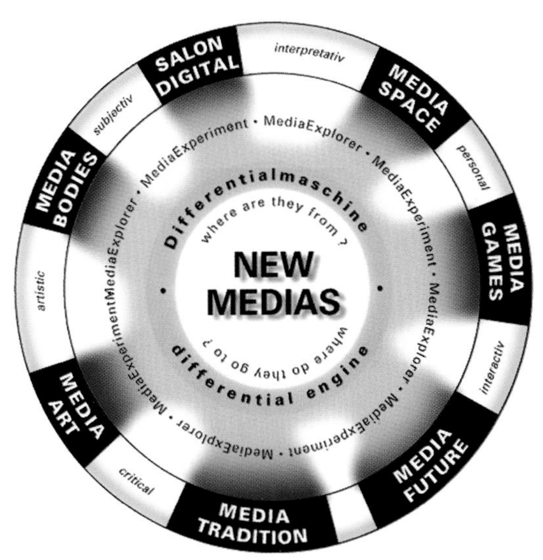

Fig. 2. MUSEOGRAMM. Graphics: Ralph Pfeifer, ZKM | Center for Art and Media Karlsruhe, Media Museum

And the instruments to answer these questions are various installations, didactic, interpretative and artificial, subjective, critical and interactive, mainly commissioned directly to artists, and build more or less exclusively for the museum.

Some examples: An installation from the Californian artist Lynn Hershman, based on the idea of Charles Babbage's differential engine is transforming the whole museum into an immersive virtual space, symbolizing the ubiquitous machine the computer that can be explored directly, through the body of the visitors.

With some experimental workstations the visitors can make their own experiments with the creative potentials of the contemporary software.

The Salon Digital will give access to the possibilities of the international network in a very different way to the traditional electronic cafés. It aims not to the well known electronic chit-chat but to global discourses prepared by a worldwide curato-in-residence program.

Some internet-coffee tables will be established in the restaurant of the center to enable the visitors to make their first experience with the communication in the global Internet.

This table found its shape through an international competition won by two young German designers.

The real virtual space of the Salon Digital will be situated in the entrance area of the museum and is dedicated to special events prepared by our curator in residence program these events could be artistic performances using the possibilities of telepresence, like the artificial games of Agnes Hegedüs or the sensual experiments created by the young English artist Paul Sermon.

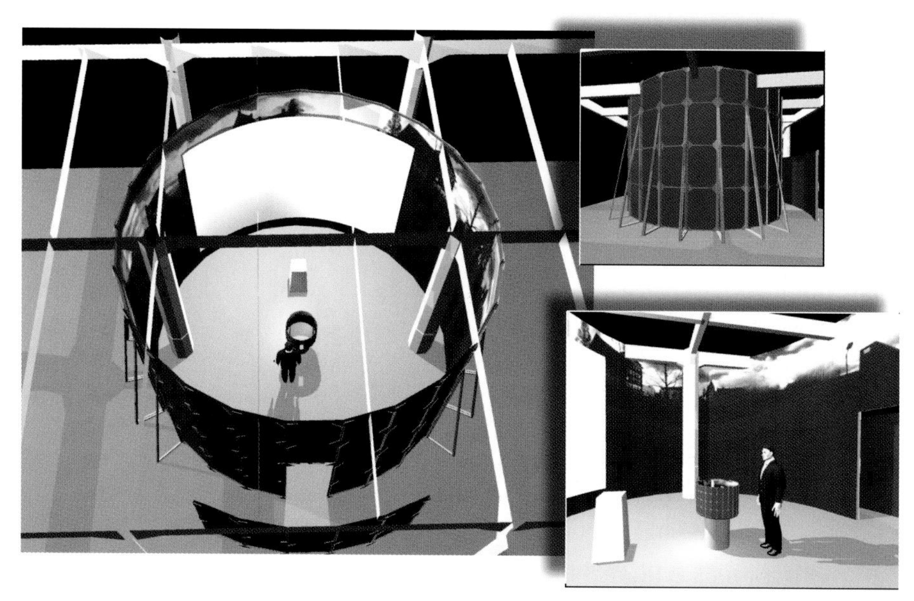

Fig. 3. Vitual Reality Theater. Jeffrey Shaw, ZKM | Center for Art and Media Karlsruhe, Media Museum

And there will be different virtual sites in the Salon Digital: one is the virtual laboratory where scientists and artists from the ZKM and from all over the world should be developing the tools to explore the Internet in a more appropriate way than it is possible until now. There will be a site to play games in the virtual media museum and to be linked with other virtual museums existing in the net.

There will be a large sculpture garden, a simulation of one part of a landscape near Karlsruhe. From this virtual landscape every artist in the world can put one piece, take it on his homeserver change it and send it back. We hope during the next years to have established a real global social sculpture in this sculpture garden.

And at last there will exist an expanded virtual architecture museum, a building designed on the rules of the network space and not the Euclidean space of our sensual experience. In this virtual architecture museum there will be rooms for virtual exhibitions of any kind and an interactive café where worldwide discussions about the future network architecture could be held.

Woven into these three networks the pedagogical network of the media experiments, the symbolical network of the differential engine and the intellectual network of the Salon Digital are the six departments of the museum with their interactive installations dedicated to the contents of Media and Body, Media and Space, Media and Tradition, Media and Future, Media and Games and finally Media and the Arts.

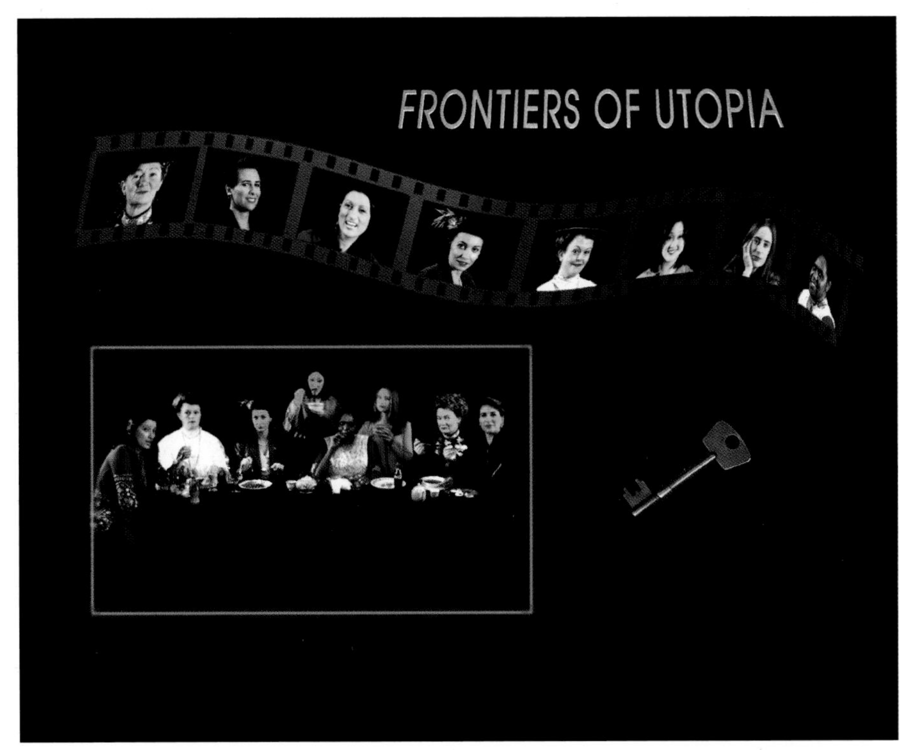

Fig. 4. Menue of Frontiers of Utopia. Jill Scott, ZKM | Center for Art and Media Karlsruhe, Media MuseumZKM

Here we will establish perhaps the first gallery of interactive media art which contains the masterpieces of this very young genre – starting with Lynn Hershman's Lorna, the possible first work of interactive media art in the world. The purpose of this gallery should be to give an overview on the many different ways which artists went in the last fifteen years to develop a specific aestethic language for interactivity at least, because this avant-garde art – as I mentioned before – until now is more or less homeless in the institutional system of the art eking out their existence at festivals, congresses or even at the edge of trade fairs. And I suppose this is the reason why the public until now has no opportunity to explore the richness of this art and can't relate it to other artistic experiments of our days.

I hope that this combination of a real museum dedicated to the history and the future of media and their influence to our society and the virtual museum of the Salon Digital could give an appropriate field of experience for the possible museum of the future.

From London to Nagasaki – The Roots of Interactive Art @ the Exploratorium

Peter Richards

Introduction

It has been noted that a common ground between artists and scientists is their process of exploration, experimentation, and testing of intuitions. These kinds of activities precede, and are germane to all scientific and artistic creation, and are inherent in children. This playful attitude has become a common value throughout the Exploratorium, a museum of science, art, and human perception, providing a unifying thread for creating meaningful and enjoyable experiences for the public. In turn, the museum seeks to work with artists who share this deep-seated joy of learning about the world, and whose investigative approaches lead not only to new discoveries and understandings, but also celebrates one of the most human of activities, the process of drawing meaning from experience.

Starting in London, we will look at "Cybernetic Serendipity" an exhibition sponsored by the Institute of Contemporary Art, then we will review examples of work commissioned through the years by the Exploratorium, concluding with a recent project "Remembering Nagasaki" a web-site based on the photographs of Yosuke Yamahata.

The Exploratorium

The Exploratorium is a museum of science, art and human perception, housing more than 600 hundred interactive exhibits and artworks that provide learning experiences about natural phenomena and human perception. It is a creative environment where visitors can explore on their own terms, linger where they choose, and experiment as they please. Perception is the underlying theme because how we see, hear, feel, smell, and otherwise sense the world, determines much of what we know about it. The tools available to visitors include scientific instruments, experiments, mathematics, language, music, media and art, creating a perfect place to make discoveries and to realize new ideas.

The Exploratorium was founded by Frank Oppenheimer in 1969. He envisioned a place that would encourage people to explore, experiment, observe and to create their own understandings about nature. He knew that the best way to know more than just textbook definitions of something is to learn through experiences. As a high school science teacher in the 50's he became disenchanted with the way science was being taught, and began inventing new ways of engaging students. A few years later, at the University of Colorado, he developed a physics class built around a number of his table-top physics experiments. Several of these were to become some of the first interactive exhibits at the museum.

During the Exploratorium's incubation period, Frank visited a lot of museums, both in the states and in Europe, and came to recognize that museums have an important role in society as places that nurture informal learning. He found that museums can re-establish an individuals' confidence in his or her own ability to understand and learn, give people a sense of their connection with the past, and a sense that human acts and human beings are a part of nature. He felt that museums could play an important role in nurturing a society of life-long learners.[1]

Art and Science Since the Beginning

Frank's intuition told him that art could be an important part of the Exploratorium. He believed that the observations and discoveries artists make could be an important aspect of his new museum. He wanted a place that provided multiple examples and multiple view-points of a particular phenomena or experience, and realized that he could look to both artists and scientists, to help reveal aspects of nature that people tend to ignore, or had never been encouraged to see.

During that early period, as we began building interactive exhibits, we came to recognize the powerful nature of this kind of communication. We quickly learned that the best exhibits were ones that incorporated beauty, evoked questions, and provided numerous ways that the visitor could engage – on a personal level – with an exhibit's inherent concept. There was something else too; an exhibit should possess a playful, open-ended quality that allowed people to follow their own nose rather than being led through a rigid, proscribed set of directions leading to the correct answer. In fact, it became part of the culture that there were no "right" answers at the Exploratorium. We came to respect and appreciate the fact that we are all creators of our own worlds; each of us develops a personal understanding of how the world works. As life-long learners, our perceptions or understandings evolve as we gain new insights and experiences.

London

Presented in 1970, "Cybernetic Serendipity" was, in retrospect, a touchstone exhibition for the Exploratorium. Curated by Jaisa Reichardt for the Institute of Contemporary Art in London in 1968, it traveled to the U.S. the following year. A major portion of the show was donated to the Exploratorium as a fortuitous alternative to its being completely dismantled at the end of its tour.

This exhibition was an exploration and demonstration of some of the relationships between technology and creativity and focused on the computer as a creative tool. It was intended to present an area of activity which manifests artists' involvement with science, and to show the links between the random systems

[1] The Exploratorium, Special Issue, March 1985

employed by artists, composers and poets, and those involved with the making and use of cybernetic devices. This show was comprised of computer generated graphics, films, music, poems, and text. These were cybernetic devices presented as works of art, environments, robots and painting machines.[2]

In her introduction in the exhibition's catalog, Reichardt states: "new media inevitably alters the shape of art, the characteristics of music, and the content of poetry. New possibilities extend the range of expression of those creative people whom we identify as painters, filmmakers, composers, and poets. It is very rare, that new media, and new systems should bring in their wake, new people to become involved in creative activity. This has happened with the advent of computers. The engineers – for whom the graphic driven by computer, represent nothing more than a means to solving certain problems visually – have occasionally become so interested in the possibilities of this visual output, that they started making drawings which bear no practical applications, and for which the only real motives are the desire to explore, and the sheer pleasure of seeing a drawing materialize. Thus people, who would never have put pencil to paper, or brush to canvas, have started making images, both still and animated, which approximate and often look identical to what we call "art" and put in public galleries."[3]

This was, for her, the single most profound revelation of the exhibition. It had a profound effect on us as well. The art work in the show shared many of the qualities of the Exploratorium's new exhibits. They were interactive, whimsical, and most importantly, suggested that there was a significant role that art could play in a science museum setting. Even though Reichardt made a conscious decision to not label the work as being done by an artist or by a computer scientist, there were a number of significant names representing the then experimental edge of art; these included Frank Malina, founder of the journal Leonardo, Wen Ying Tsai, Jean Tinquely, John Cage, James Tenny, Karlheinz Stockhausen and Nam June Paik.[4] In retrospect, the works, representing a dialog between artists and technology, were bellwethers of what was to follow, nationally, internationally, and particularly at the Exploratorium. As the first artworks presented by the Exploratorium, they provided an alternative view of new media's creative potential.

It was not long before the local art community began to see the Exploratorium as a place where highly experimental work, particularly work involving technology, was being supported. The local chapter of Experiments in Art and Technology (EAT) did several sound performances in the 120,000 square foot open space in 1969 and 1970. During this same time, Bob Miller brought in an idea for creating a real-time painting using refracted light of the sun for his pigment. Experimenting with large prisms picked up in an army surplus store, Bob found that he could create large washes of color by using these prisms and crumpled mylar to reflect refracted light onto a projection screen. An extended period of

---

[2] Cybernetic Serendipity: The Computer and the Arts, p. 5

[3] Cybernetic Serendipity, ibid.

[4] Cybernetic Serendipity, ibid.

research and development resulted in an installation that invites simple investigations into the nature of light, in a highly theatrical setting.

Sun Painting by Bob Miller
Twenty-seven years later, this work remains one of the most successful ever created at the museum by accomplishing the difficult task of being both a model science exhibit, and a consummate work of art.[5]

Fig. 1. Sun Painting by Bob Miller, Exploratorium. © San Francisco 1970

Formalizing a Program for Artists
A few years after the Sun Painting was completed and upon the receipt of a National Endowment for the Arts grant, the Artist-In-Residence Program was formally organized to actively seek and commission artists to create new works for the exhibit floor. As this program gained momentum, the Exploratorium began experimenting with presenting works by musicians, performance and theater

---

[5] A Curious Alliance: The Role of Art in a Science Museum, p. 4

artists, video, film and craft artists. The inclusion of these disciplines into a science-based culture spawned and continues to generate long discussions about the relationship between art and science, and the role that these two different ways of looking at the world can play in helping people develop an understanding of their own place in the world.

We have also come to recognize that the roads of understanding and learning are different for each individual, so we furnish a multiplicity of experiences, which taken together, offer a matrix of learning paths for people from a wide variety of backgrounds and sensibilities to find out about themselves in relation to the world. The Arts Programs provides multiple windows, doors, and even skylights, illuminating and providing access to aspects of nature and society that are not always attainable through other museum mediums.

Physical phenomena has provided a point of departure for many artists working at the Exploratorium, including, Miller's "Sun Painting" now considered by many, to be the museum's quintessential work. The development of Michael Brown's "Meanderings" followed a similar, rather circuitous route to realization. Armed, with an excitement about the way water trickles across a sheet of glass, with limited technical skills, and with no experience in producing something to withstand the rigorous demands of a museum public, Michael had to quickly learn and utilize the collaborative culture of the Exploratorium. He also had to learn how to gracefully receive and filter the numerous suggestions that came his way everyday from the museum's shop staff.

> "I started with a vertical piece of glass, then people in the shop came over and kicked around ideas. What happens if you tilt it? What happens if you use lights? What shadows does it make? We discovered, among other things, that if you placed a piece of light-colored material below the surface of the glass and added a pinpoint light source, the water acted as a lens. We played with it and got these beautiful optical effects."[6]

Meandering: by Michael Brown

Michael's experience developing "Meanderings" became a tricky balancing act of receiving advise, of considering each suggestion, and at the same time, of trying to hang onto the essence of his idea so that the piece did not become someone else's. This sort of feedback was interspersed with an on-going dialogue with staff scientists and other exhibit developers interested and excited by the new things they were seeing in Michael's work. As the work developed, each person shared their own observations and theories of what they thought was happening, creating a charged atmosphere that extended Michael's understanding of resistance and cohesion in water, into many other realms. Now that the work is out on the floor, it continues to provoke conversations, play, and speculation about the nature of water.[7]

---

[6] A Curious Alliance: The Role of Art in a Science Museum, p. 19

[7] ibid.

Fig. 2. Meanderings by Michael Brown, Exploratorium. © San Francisco 1993

A different kind of investigation and a different kind of atmosphere was created by the recent presentation of eight performances, staged simultaneously, by performing artists who based their works on the way Exploratorium visitors use the museum space or its exhibits. These performances reflected the theatrical aspects of a museum that encourages interactions between people, people and exhibits, and between staff and visitors.

Performing artists, working at the Exploratorium, are encouraged to examine culture as a perceptual phenomenon. The common bonds these artists share are an innovative approach to their art form, and a desire to explore technical and creative problems in the development of their work. Performance at the Exploratorium involves people on all levels; as creators, performers and audience, creating the ultimate in social experiences – holding the mirror up so we can see ourselves, both as individuals and as part of a cultural milieu. This particular project is a component of a museum-wide assessment of how well the museum functions, both operationally and as a community educational resource, after being open for more than twenty-five years. This process will determine how the museum will manifest itself in the future.

Dr. Itsuo Sakane, journalist, curator, and President of The International Academy of Media Arts and Sciences in Gifu, has been a mentor, a wonderful source of inspiration for the Arts Programs at the Exploratorium, and a key figure in our international network of people who suggest new artists for our programs. Several years ago he recommended that we invite a brilliant young artist, Toshio Iwai, to be an artist-in-residence.

Using modern technology, Toshio's early work investigated the potential of advancing 19th century moving-image techniques, such as flip books, zoetropes, and kinescopes. His first installation for the Exploratorium "Well of Lights" was a manifestation of his interest in transferring moving image ideas to three dimensions. It goes well beyond film, and the video-monitor or television format; considering some revolutionary ways in which an object and image can appear to move in space. As ethereal as "Sun Painting" this image sculpture generates six layers of stacked moving images, using a combination of computers, strobing video projections and spinning transparent discs.

Musical Insects: by Toshio Iwai

Toshio's second work is one of the most popular pieces in the museum. "Musical Insects" enables people to compose music using a graphical paint system he developed for an Amiga computer. Little points of colored light (insects) run around on the screen of a monitor, each making a different sound – piano, bass, or percus-

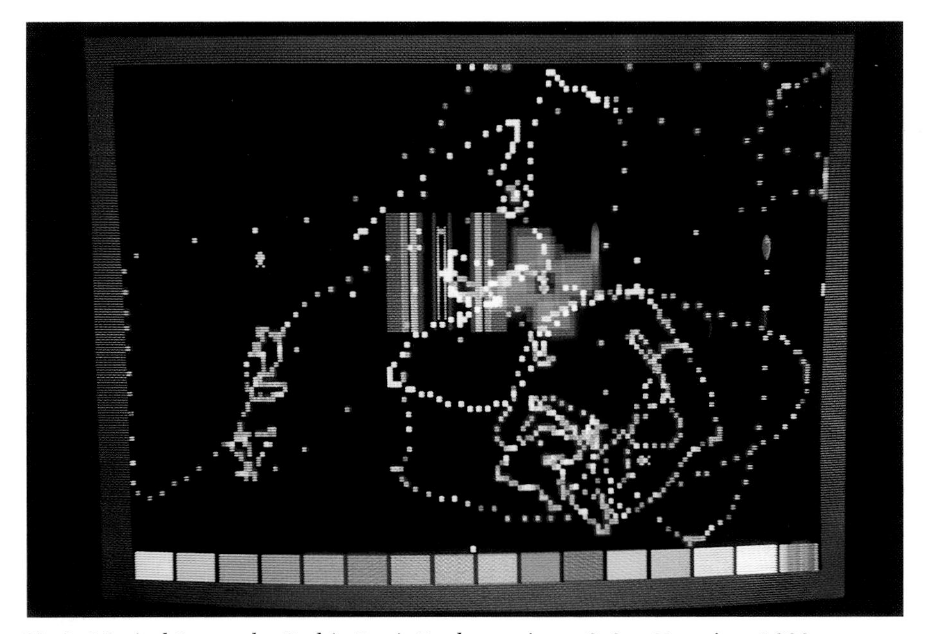

Fig.3. Musical Insects by Toshio Iwai, Exploratorium. © San Francisco 1992

sion – when they hit another point of colored light on the monitor. By painting splashes of color in their paths, musical tones are created when both intersect. A specific color generates a specific tone. Some colors also cause the "insects" to change direction, providing opportunities for creating rhythmic relationships. Maxis Inc., a computer game company, working in collaboration with Toshio, has recently published a CD ROM version of "Musical Insects" for home computers, entitled "Sim Tunes."

Nagasaki

In conjunction with "Nagasaki Journey" the simultaneous exhibition of Yosuke Yamahata's photographs in Tokyo, New York and San Francisco, we were invited to create a media experience that would extend the reach of this important presentation. At the same time, because we were starting to work on a new exhibit section on memory, we saw this project as an opportunity to experiment with the relationship between personal memories and broad historical events, and to take part in remembering an event that has shaped the thinking and sense of future of all peoples on this planet.

Mr. Yamahata, a young photographer, entered Nagasaki the day after the

Fig. 4. Remembering Nagasaki, Exploratorium. © San Francisco 1996. Photographer: Yosuke Yamahata, Nagasaki 1945.08.10

atomic bomb was dropped by U.S. forces on August 9, 1945. His record of the devastation, originally commissioned by the Japanese government, was kept secret and not seen by the public until his family released his photographs for display as part of an international commemoration of this event fifty years later. Three artists on our staff, Alison Sant, Marina McDougall, and Susan Schwartzenberg, worked with a team who was developing a new exhibit section on memory. Using Mr. Yamahata's photographs as a point of focus, they created a web-site for visitors to explore the relationship between their personal memories and larger historical memories surrounding atomic warfare, and provided a context to surface their own memories about atomic warfare.

Remembering Nagasaki

For many, "Remembering Nagasaki"[8] became a digital journey; one that wound itself through the devastation of a city; each place and event captured by the unforgiving eye of black and white photography. The journey was extended to the recorded thoughts, opinions and remembrances added to the site's commentary pages by other visitors to the web-site, including those who were in Nagasaki on August 9, 1945, or those whose lives were affected by the events of that time. Site visitors had options to view the photographs, or visit its three other pages "Atomic Memories" "Commentary" or "Commemoration."

We were particularly interested in how people gather and alter stories about events they did not directly experience. "Atomic Memories" provided a place for people to record their own recollections of how they learned about the bombing of Hiroshima and Nagasaki. "Commentary" invited visitors to submit their ideas and opinions of the bombing, nuclear war, and living in the nuclear age. This section became a forum for people to engage in a global conversation about these and other issues, including the process of writing history, the ethical responsibility of scientists and technologists, and the decision to use the bomb. "Commemoration" provided a list of related films, videos, CD ROM's, and books, as well as suggested links with other sites that contained current information about Nagasaki and Hiroshima, eye witness accounts, archives, papers and other materials related to nuclear science and the atomic age. Thousands of responses came in, and were posted by the site's creators, from all over the world.

The artists who developed "Remembering Nagasaki"[9] had a sense of what they were creating, but were unprepared for the tremendous response that their work generated.[10] Watching people use it as a tool, we began to get a picture of how

---

[8] http://www.exploratorium.edu/nagasaki

[9] Although no longer active at the time of this writing, the site can be accessed as a document http://www.exploratorium.edu/nagasaki/mainn.html

[10] The technology for, and thinking about how to create threaded discussions on the web had not yet caught up with peoples desire to participate. No software was readily available at the time to make threaded discussions on the web, so each of the entries had to be re-entered by the artists, into the site by hand.
A Curious Alliance, p. 21

people learn about events – using each others stories and perspectives to fill in their own informational gaps.

The site became a place where all concerned were learning about history, nuclear war, science, social issues; and learning about these things from multiple perspectives. At the same time "Remembering Nagasaki" provided a place for people to express, in emotional terms, their feelings about global situations that had deep personal meaning for them, situations which they had no direct control over. By participating in what I see as the act of creating a history, these people became a part of that history, and that history became a part of them.

Conclusion

When Michael Brown finally placed "Meanderings" on the exhibit floor, he thought the dialog had come to an end – the one that transpired between himself, the Exploratorium staff who helped him develop the piece, and on occasion, the public who gave him feedback at various stages. He found out differently:

> "I'd worked with it for months and months and I thought I knew everything it could do. But within a week, people on the floor were playing with it in ways that never even occurred to me. That's the way things work here. You create something, and even when you think it is done, it keeps going from there. I made it originally, so I can claim it as mine, but other people are continually re-inventing what it can do."

The surprise was that "Meanderings" became much more than just transferring information – it became a conversation in a way – what it really became was a collaboration. The people who are now using the piece on the museum floor are continuing the process of discovery, far beyond Michael's vision for the piece. Like the engineers at Bell Labs, who began drawing with computers, the novice composers creating music with "Musical Insects" or the people who contributed their thoughts to "Remembering Nagasaki," they all became caught up in something much larger than themselves even as they took ownership of it. This is the nature of true interactivity, and is what we strive to achieve in our work with artists and scientists at the Exploratorium.

References

Cybernetic Serendipity: The Computer and the Arts, Studio International, 1968, W. and J. Makay & Co. Ltd., Chatham, Kent, U.K.

The Exploratorium Quarterly, Frank Oppenheimer, Special Issue, March 1985, The Exploratorium, 3601 Lyon Street, San Francisco, CA, U.S.

A Curious Alliance: The Role of Art in a Science Museum, 1994, ISBN 94-3451-39-6, The Exploratorium

Remembering Nagasaki: http://www.exploratorium.edu/nagasaki

7. Education of Art & Science

The Historical Background of Science-Art and Its Potential Future Impact

Itsuo Sakane

I. Human Endeavor in Making a Bridge Between Art and Science

Why Science-Art?

For people who have witnessed the development of diversified types of contemporary art since the 1960s, the category of "Science-Art" may seem a little bit unusual. So-called Technological Art has been well accepted as an established art category, and new names such as Media Art, Cyber-Art, or Web-Art (or Internet-Art) are becoming commonly used and referred to today. The tendency to create a new category is becoming more-or-less a socio-cultural phenomenon.

This new name connecting the two extremes of culture, Science and Art, contains a different nuance than other new art forms which have been conceived and developed primarily using technological breakthroughs or new industrial materials. When people hear of an art-form called "Science-Art" they may possibly think that there is a complex context behind the name. It might sound like just Science and Art or even the concept of Art with Science. Frankly I, myself, have been using this new category of "Science-Art" in this context for the past ten years.

I created an introductory course called "Science-Art" at the Shonan Fujisawa Campus of Keio University in 1990, where I taught until the end of March, 1996. I also directed and curated an exhibition at the Science-Art Gallery for the Japan Pavilion at Expo '92, in Seville, Spain. The reason I did not use the names "Electronic Art" or "Technological Art" was that I felt they were not comprehensive enough to describe the deep meaning of these new art forms wthin the context of Art and Science today.

These are not only my thoughts and opinions. Mr. Frank Popper, a French critic in Paris, stated that the definition of Technological Art is no longer appropriate today, and he suggested a new name for this genre should be "Techno-Science-Art". It sounds much more practical and concrete to express the meaning of art in the context of Art, Science and Technology. However, I have the strong feeling that we should not only include new technological artwork, but also new artforms which have been made possible by the introduction of a new "world vision". This includes knowledge gained from observing nature and the universe gained from new scientific discoveries since the last century. This is more than just the application of new scientific technology, but is based on a new way of looking at the time-space concept, a new cosmic view, and a new view of nature which is being influenced by a new scientific concept today.

The Socio-Cultural Background of Science-Art

Many new types of categories of art have been established since the 1960's, such as the genre of, say, Kinetic Art, Optical Art, and various Media Arts, as well as Land Art and Cosmic Art, etc. These trends are similar to the history of science in the development of newer categories of research. It seems rather a paradox for the art world, though, because art itself has historically been appreciated by human beings as a whole entity from the beginning of history. For that reason, an endeavor to recover the wholeness of art from scientific reductionism, which tends to make human activity more mechanical and inhumane, has been gradually appealing to thoughtful people.

C. P. Snow's "Two Cultures" (1958), and Gyorgy Kepes's "New Landscape"[1], are both examples of this expression of consciousness among people whose desire was to re-unite the split Science and Art world of that time. The new movement of Art and Technology, which began around the world at the end of the 1960's, was the result of a chain-reaction of this new consciousness.

I, myself, am reminded of three exhibitions I saw at the beginning of 1970s in the U.S. which reflected this new consciousness; "Art and Technology" at the County Museum in Los Angeles, "Software" at the Jewish Museum in New York, and the "Four Element Show" at the Boston Museum of Art. Those three exhbitions were the result of consciously challenging the human mind to bring together an aesthetic sensibility and scientific innovation even though their stand points for the show were somewhat different according to each exhibition.

It is not only a simple expression of joy to use the new methodology or new materials from this new technology. In the the newly emerging art forms of cosmic arts or natural phenomena arts, there is a strong, conscientious effort being made to make whole the formerly split division between science and art. An example can be seen in James Turrel's "Roden Crater Project", or Charles Ross's "Star Axis Project" in New Mexico, U.S. Both fall within the new category of Science-Art rather than Technological Art.

In the category of Science-Art, we can include the various technological art trends which have appeared since the 1950's as a dynamic whole structure. This trend will continue for future generations.

The Status of Science-Art Today

Looking back at the various exhibitions, events or symposia which have taken place all over the world during the past 15 years, we can see that there has been a strong movement to unite art and science. This movement has been proceeding steadily and is becoming more and more visible. There was satirical criticism directed towards the techno-scientific civilization seen in the exhibition of 'Immaterial' at the Pompidou Center in Paris in the 1980's. At the Venice Biennale'86, Artwork, in the theme 'Science and Art', we could see this broad new scope in numerous movements from archemistic art to technological art, to biological art and to the development of communication art as a symbol and representation of human activity in the future. The rooms for displaying the artwork was arranged in a mosaic style.

The reason why I labeled the exhibition I organized for the Japan Pavilion at Expo'92 as the "Science-Art Gallery" was that I could not neglect the trend of consciousness prevalent at that time, even though I introduced some technological art forms in the show. I daringly introduced the metaphor of the Japanese tradition of the "Summer Festival Garden" at this art show because I felt that introducing an aspect of Japanese culture might compensate for the universality of the scientific concept[2].

I have recently found similar themes of Science-Art in many symposia being held around the world. The ART-Science-ATR conference, which was held in Kyoto in 1976, was just one example.

If we look back at the history of the journal LEONARDO, which was founded by the scientist and artist, Frank Malina, who incidentally was also one of the founders of the Jet Propulsion Laboratory at NASA, the journal's aim was the integration of Art, Science and Technology. The current Executive Editor, George Malina, son of Frank, is also a scientist and is following in the footsteps of his father. The LEONARDO's home page on the Internet was selected recently as one of the best three in the field of art and science home pages. Additionally, the participation in Science and Art by industry has been increasing more and more. The Louis Vitton Company in Paris has developed an award system for Science for Art, and even the other French company, Loreal, is in the process of creating another type of award system for contributors to Science-Art.

In Japan, there are some salons for scientists and artists which have been active, such as "Japan Society of Science on Form" or "ARS+". In reference to the latter, the famous Physicist and Emeritus Professor, Kodi Hushimi, worked as the honorary chairman. The "ARS+" was disassembled after organizing an international conference of "Katachi∪Symmetry" last year. Since then, Prof. Ryuji Takaki, also a physicist, has been trying to organize another salon for scientists and artists in Japan.

Our new school, IAMAS, which opened in Ogaki city, Gifu Prefecture, in April last year, 1996, is called the International Academy of Media Arts and Sciences because its aim is to integrate information science and art in the future. In fact, there are many research institutions and universities in the world wherein scientists and artists are collaborating. At the Super Computing Center at Illinois University, at Urbana-Champaign, for example, an artist, Donna J. Cox, has been working with other computer scientists to make a Scientific Visualization for increasing aesthetic appeal. At the Xerox PARK Institute and at Interval Reseach in Palo Alto, California, there are artists-in-residence working together with computer scientists, and this liaison is designed to stimulate their imagination. In the field of mathematics, one of the historical fields of Science-Art, a great deal of mathematical art has been created, and it has become a rich source for artwork.

Now, particularly in this age, much artwork in the genre of Science-Art, has been connected to each other because of the use of computer media and the connection to network systems. This has been evolving into familiar styles of visible artwork for the public, from child to adult. Software can be used for edutainment purposes, wherein difficult subjects like mathematics, physics, or even biology can be taught. This very enjoyable, self-educating lesson software has been increasing and developing as a new field.

II. Recovering the Wholeness in the Information Age

Conflict Between Material and Immaterial

But on the other hand, we have been more and more sensitive on the paradox within the information age, where we are gradually forgetting about the real dialogue with substances, inner self and the surrounding real world. Especially the appearance of the virtual reality technology in the past decade has been deeply influencing to such technoscientific art today and the earlier type of art based on the material or natural, physical phenomena has been gradually out of trend. Art is more or less tends to be immaterialized, and even the sensual experience here is becoming the pseudoscopic one. It is a paradox for the human evolution that the more knowledge of information and more global consciousness mankind develops, the more his/her concern tends to go out from the daily personal sensual level toward the intangible reality lacking in the real contact through senses.

But if we look back, we see that art itself started as a way for the mind, each person's inner self, to experience virtual reality in a wider sense. From cave paintings, through the writing of fiction, to the miscellaneous forms of visual art today, art has always worked as a catalyst to evoke in the imagination each person's sense of reality. It is the imagination itself, however directly or indirectly, that must arouse in each individual memory the sensual experiences of one's daily life. Marcel Proust was able to revive memories of times past just by dipping his Madelaine tea-cake into a teacup: the tactile and olfactory senses are able to act as a catalyst to inspire this imagination.

When we compare such vivid sensual experience of daily life in times past to the urban life of today's consumer, automation technology, we often feel that we have lost contact with the vividness of nature. The recent appearance of virtual-reality art might, in a sense, be running parallel to the modern socio-psychological human condition.

Although it is no doubt a great challenge and a great act of creation to develop the technology to stimulate a realistic relationship between people and environment, one has to wonder why artists, who are likely to be among the more insightful and sensitive of human beings, must confine their activities to a narrow monologue within virtual-reality digital technology.

We need to be more concerned with the role of art as it affects all of us. Art has, of course, always been humanity's highest form of the expression of freedom and imagination. But, at the same time, art might well have started, as Jacob Bronowski suggested, as a tool for human survival. We have relied on the compensating power of art for centuries. Since the Renaissance, we have seen art not merely as a kind of 'sleeping pill' but also as something with genuine healing power, a stimulant to recovery. It is in such society as ours, in which one tends to be unable to distinguish between reality and artificial reality and ends up chasing daydream, that the healing power of art is most effective.

We understand now what Joseph Beuys wanted to express by using raw materials in his artworks. He used organic and inorganic substances such as fat, felt, beeswax and copper metal in his works not only as a metaphor for his own critical experiences in wartime but also as a form for social and ecological art. His inner motivation was evident in his active membership in the Green Party.

I am not at all against the art of virtual reality. I value it highly and believe in it, frankly, as a way to connect our minds to our environment. Many friends artists in this field have been trying to use virtual-reality media not to escape reality but rather to gaze at it more directly. Nor do I overlook the positive role that information media technology such as internet communication media have been played recently in helping generate global consensus on environmental problems and issues of human rights.

Let me make clear at this point that what I am arguing for is a balance between both types of art in this age. A harmony is possible in our personal and social lives between the art of synthetic reality, based on information technology, and the art of the five human senses, based on material.

Challenge to Recover the Wholeness

As I already mentioned before, I have a vivid memory of the exhibition of "Earth, Air, Fire, Water: Elements of Art" at the Museum of Fine Arts in Boston, in 1970. It was an eye-opening show for the public: the flow of liquid, the power of surface tension in the soap bubble or soap membrane, hidden solar energy and the behavior of natural phenomena. Those images were really inspired with the recent scientific perspective as Gyorgy Kepes introduced in his "New Landscape". A similar approach to art can be traced back to Goethe or Rudolf Steiner and is more closely related to the ecological problems we face today. It was also reflected in the adventure-filled Exploratorium as a place for everybody to participate with their own five senses in the wonders of art and science.

On my way back to Japan after spending one year as a Nieman Fellow at Harvard, I visited the University of Utah in 1971 and experienced a very innovative, head-mounted device containing a 3D viewing glass. This glass enabled the wearer to walk around in a wire-framed computer-graphics image floating within the room. It was actually the prototype of the 3D viewing helmet in its earliest stages. On the way from Utah, I dropped in San Francisco, to meet with a scientist at Smith Kettlewell Eye Research Institute who had been developing the tactile perception system for the blind people to use their back. I first saw his devices at the "Software" exhibition at Jewish Museum in New York in 1970. There I was also impressed with Nicholas Negroponte's "Seek", which was a simmulation model of future city for gerbils.Within a big glass box containing many building elements of blocks and many gerbils as havitants, gerbils' behavior could control the shape of their 'city' with sensing devices and computer controled robot-arm. It was really the unique exhibition where many artists and scientist gathered in the creation of new type of art. That was a time, the early 1970s, in which people participated in both types of science-art either using information technology or natural phenomena, and switched back and forth from one type to the other.

Now as Negroponte's bestseller book of "Being Digital"[3] symbolizes, people are more and more inclining to the digital technology, and seem to be forgetting about the dialogue with material and natural phenomena. I am a little bit afraid about such tendency if it would continue among the young generation. In providing art education to young children, it is urgent that we find a balance among art activities. If all they learn in school is computer literacy, or digital technology, is it really enough that we turn out computer hackers? There must be an alternative way to lead them to interactive dialogue with materials by means of their five senses.

Computer power can be connected to both real-life activity and to the creation of mixed-media sculpture and both kinetic and phenomena art. Group training could be conducted to improve the chances of children making contact with the natural environment more often. Ten years ago, when I met Ann Marion at Vibarium Project in Los Angeles, within this project children participated not only in creating the computer simulation of the fish movement but also experience in the real sea water to interact with the behavior of the fishes and plants within the real ecological environment. It would be working as the real healing factor toward the "Virtual Reality Syndrome".

Art for Survival

Thus it would be the more important for the artists and for every person who wishes to share the new experiences of the art to let both material and immaterial type of art, or both sensual contact and information type of art coexist and fuse together for the future survival. It is the proposal for the two art problem which appeared now after the two culture problem between science and art in 50's.

When I visited David Bohm at his Birkbeck College in London in 1978, he told me that the art had been the basic form of the humanistic expression of the wholeness comparing to science, and even the science used to be one form of art. According to him, science had been so much specialized and become the reductionistic discipline and so he wished the science to be back to the similar discipline as art. Comparing to such trend of sciences, I felt the similar reductionistic style had been overwhelming in recent art. If the main trend in art today continues to follow its present situation, it might be reductionist and lack a wholeness, too.

There is some hopeful sign for future, though. In the past few years, the new trial among artists/scientists in developing the new human interface to make a bridge between such virtual and real world had been more active. Especially there are appearing some interesting art forms using such sensual interface in the field of Interactive Computer Art. They could be the new catalyst for people to understand the recent view on the world, from micro to macro, which were based on the new scientific discovery, and would induce to the rediscovering the wholeness. We would be able to come closer even to the insightful scope on the inter-relationship between the brain and its consciousness in future.

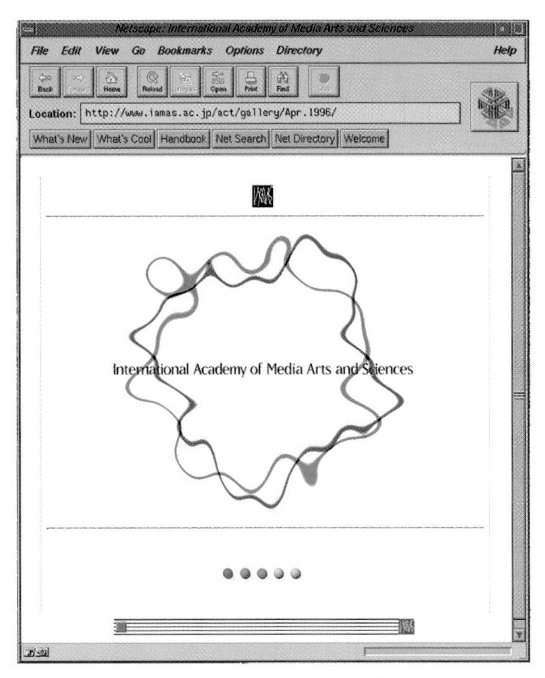

Fig. 1. IAMAS home page. © Academy of Media Arts and Sciences, Gifu Japan

Fortunately, I could have chances to organize two different types of exhibitions within this year; one as "Interaction '97" held in March, 1997 at Softopia Japan Center in Ogaki City, Gifu Prefecture, where we invited 10 interactive installation art works from all over the world, and the second as "Art and Science" exhibition held at ICC museum in Tokyo from July 29 till September 7, 1997. Especially in this Art and Science show, I could organize the phenomena type of art works under the title of "Sensitive Chaos" for which I selected about 13 art works mainly based on the natural phenomena like liquid movement, vibration, surface tension, reflecting light, wave phenomena, gravitation, crystallization, and even the cosmic hidden pattern between Sun and Earth. It was intently done to keep the sense of balance toward recent media art movement which are more or less consisted with digital, and computer oriented art works. At this show, many audiences were strongly impressed in such a chance to be able to have an inner dialogue or meditation with those phenomena's unimaginable beautiful behavior. It was really beyond the classification between art and science, and directly appealed to the innocent human minds from children to aged people. Many visitors confessed even to have had a feeling of relaxtion and recovered the freshness from those expriences.

Epilogue

I apologize for writing this general discussion in rather ambiguous essay style including my personal experiences since late 60's. I hope you could understand what I had been trying to make a bridge between Science and Art for more than 30 years, sometimes as a journalist[4], sometimes as a teacher and lecturer in a college, and sometimes as a director/curator of many science-art exhibitions[5]. Fortunately I could have so many chances to meet both unique artists and scientists in the world who had been sharing the sympathy in the integration of art and science. Charles and Ray Eames with whom I was first acquainted in 1967 were such influencing artists. Frank Oppenheimer, Piet Hein, M.C. Escher, Stanley Cyril Smith, Frank Malina were also those influencing geniuses who unfortunately have passed away recently or in the last decades. I also have been inspired from the encounter with Gyorgy Kepes, Martin Gardner, Kodi Hushimi and many younger friends artists and scientists in the field of border area. They have been also sharing the same spirit and expanding the chain-reaction of the creative ideas among the young people. To realize the real integration of art and science in future, I think it is the most important thing to continue such chain reaction of inspiration among young people by giving them the truly insightful concepts of science-art. In our IAMAS, I have been trying my best to stimulate our students in discovering such wonderful world by themselves and open their consciousness toward the creation.

References

1 Kepes G (1966) The new landscape in art and science. Bijutsu Shuppansha, Tokyo (Japanese Edition)
2 Sakane I (1992) Summer festival in the light garden in Japan: The Science Art Gallery at the Japan Pavilion Expo '92. LEONARDO CURRENTS No. 4
3 Negroponte N (1995) Being digital. ASCII , Tokyo (Japanese Edition)
4 Sakane I (1991) Recovering the wholeness of art: information versus material. LEONARDO, Vol.24, No. 3: 259–261
5 Sakane I (1996) Some introductory comments for the exhibition on light arts. In: Catalogue of Primitive Scene of Light, SCIENCE-ART – Invitation to Space of Future, Ibaraki, Tsukuba Museum of Art

The Evolution of Images Between Chaos, Art and New Media

Michael Klein

Introduction

The INM – Institute for New Media Frankfurt defines itself as a forum for artistic, scientific and applied questions of New Media. It deliberately assembles artist, scientist, philosophers and information scientists at a hot spot of space time to explore virtual spaces in theory and practice. In order to explain why New Media demands for this multi-disciplinary examination (the networking community(V)) the first four chapters about images and vision of cosmos (I), images and movement (II), images and interaction (III) as well as image and experience (IV) will give some motivation.

What is the discussion of New Media all about? Actually it is only about images but not images understood as visual effects but as visions of cosmos, images not seen as simple mappings of reality but as dynamic processes, images not as static frames but as interactive spaces and images not only seen as image worlds but as world experiences.

I. Images and Visions of Cosmos

In the following historical summary I will try to collect some steps of the evolution of visions of cosmos – certainly a very subjective one – which are directly or indirectly relevant for the actual discussion about concepts of New Media as well as for its inter-disciplinary meaning. The design of visions of cosmos is accompanied by mankind's attempt to survive in a complex environment. Visions of cosmos do not by chance base on image worlds. Seen in an evolution biological way, an explanation for the success of the species Homo Sapiens could be the ability to interpret images on which our space-time orientation as well as our brain function is founded.

If we are looking for a possible scheme to understand the evolution of the human knowledge we will find an exciting interplay between images and visions of cosmos over millenniums. It would be enticing in this context to start with the prehistoric cave paintings which are our earliest – and absolutely virtuous – examples of figurative – symbolic views of life, but we have to withhold ourselves to do so. From our toddies point of view, we could risk an analysis but besides the paintings themselves, we do not have enough additional information about the conditions of life from our ancestors.

Fig. 1. Skylink Frankfurt is an architectural discussion of the interactive net-cities of the future realized in vrml environments of the world wide web. The skywalk is the location where all inhabitants are accessible. Artist: Bernhard Franken, Thomas Bornfleth, © INM-Institute for New Media 1996

The incomprehensibility of the cosmos and the complexity of nature our ancestors tried to face with different gods worlds. For the Western cultures the Greeks added to these creation mythologies their natural-philosophic experiences. According to Anaxagoras the first gods Gaia and Eros of the Greek myth came into being from the KAOS – the tohuwabohu which is the chaotic pre-mixture. This structure formation was stimulated by the NOUS which means the ghost as the only non-mixable entity. The dynamics of this structure generating process is known as Perichoresis. This mythology enriched by some natural-philosophic ideas was further developed by the two Greek schools of the Phytagorists and Aristotelists. The visions of cosmos following from this foundation formed the two most influential antique theories of cosmos. They mainly differed in the role given to geometry – which is according to the antique understand the visualization of mathematical world models. The Phytagorists (Phytagoras of Samos 582–497 B.C.) consider the mathematical harmony, we mean the geometric symmetry, as the final truth: The number is nature and subject of all things and the ideal of the universal mathematical harmony. The abstract image of the mathematical model or the figurative geometrical visualization, which is the same in the antique context, is in case of doubt prior to the daily experience of the common sense. Therefor there is a direct link from the Phytagorists as protagonists of the non-objective images to today's axiomatic, seemingly objective, theoretical foundation of physical science. In this context it is very interesting that Aristarch of Samos (320–250 B.C.) succeeded in computing an heliocentric model of the sun system out of the antique understanding which was surprisingly correct as we know today, but which only was rediscovered some hundred years later by Kepler.

The epistemology of Platoon (427–347 B.C.) is part of the same tradition of Phytagoras: The truth can be found in the universe of ideas, it can only be grasped by the consciousness mind. Knowledge therefore is memory and the sensory perceptions help as incomplete shadows by realizing the primary ideas hidden behind (the originals). Platon's parable of the cave gives an early formulation of the epistemological problem of our universe. We are observers of images and we are even unable to answer the most simplest question, namely if we are dealing with shadows of reality itself or with mappings of model worlds, which only pretend to be reality.

Opposing Phytagoras and Platoon the Aristotelian philosophy (Aristoteles 384–322 B.C.) claims to be a synthesis and an holistic understanding of the universe. Following from the everyday rationalism of the common sense it demands for everything, either from terrestrial or cosmic nature, its natural position within a hierarchical order respectively perfect harmony. Mathematics is the "tool" of the common sense and accordingly, geometric models are understood as useful but only pedagogical worthy descriptions without physical, say real meaning. Again there is an interesting fact in our context. The cosmological model of Ptolemaios (90–168 B.C.) formulated according to the Aristotelian spirit was a geocentric model of the solar system and therefor from nowadays point of view a wrong model. Nevertheless it maintained the most complex imagination of the cosmos and was unsurpassed in qualitative exactness until the 14th century.

Very informative for the history of arts interpretation of the human kind is the discussion about the interpretation of pictures in the Byzantine picture quarrel of the 8th and 9th century between Iconoclasts and Iconoduls as well as in the period of the picture storms of the reformation at the beginning of the 16th century. Fortunately the theoretical discussion of epistemology never adapted the vehemence of religiously motivated discussions, but even today some discussions about computer generated electronic picture worlds in the art scene as well as in the supposedly objective natural sciences remind fatally of the senseless position fights in the dark Middle Age.

Let us skip the millennium between Antiquity and Enlightenment and let us collect some younger characteristically epochs with their technically and philosophically correlating aspects in visions of cosmos. The art-engineers of the Renaissance, all above Leonardo da Vinci (1452–1519) as the godfather of the modern Media Art was the first in exploring visions of cosmos. The Age of the Mechanistic World View began with Descartes (1596–1650). This epoch developed the foundation of the modern natural scientific theories. Johannes Kepler (1571–1630) succeeded in making a synthesis of the solar models of Ptolemaios and of Aristarch. The time of the Clockwork Universe began with Isaac Newton (1642–1727) and Gottfried W. Leibniz (1646–1716) and the belief in the analytic ergo the formal calculus of the universe. The success of the analytic differential theory again lowered the meaning of geometry, and by this the image worlds, to the role of pedagogic illustration of the contexts.

It should not be forgotten here that the mapping of reality is not unknown, especially not to the natural scientific theories. The opposite is true, the epistemic-theoretical approach in generating scientific models is of central importance in our context because from my point of view it is the basic of all reflections about virtual words and their interpretations. On the other hand there is nature or our real environment whose inner structure we would like to understand.

The generation of abstract models of the classical scientific theories means the idealized mapping of the reality in abstract, as a rule mathematical state spaces. These simplified models which contain the most basic features of reality are simulated with mathematical analysis. One hopes that the knowledge which is gained from these mathematical models will at the same time give insight in the contexts of reality. Especially the axiomatic theories which are considered as objective demand for this ideally one-to-one mapping between model worlds and reality. The experiences of our century prove the triumphal successes of this theoretical approach as well as fundamental compatibility problems among the different scientific approaches. Einstein's Theory of Relativity unified the imagination of space and time under a continuum. Cause-effect-relations became a phenomenon of a spacio-temporal local observer. The Theory of Relativity can be analytically formulated in extent but Einstein himself describes in his "thought experiments" his own basic geometrical and visual reflections which stand before any analytical formulation. (How does the world look like from a Photon's point of view?).

The Quantum Mechanics did not only become the best formulated scientific theory until the middle of our century but also presented the holistic vision of the world with the possibility that everything is connected with everything. But the Quantum theories were seen as formal theories without any chance of visualization for a long time. No images which would interpret the real processes on an atomic level were known. But meanwhile there are geometric and visual models which illustrate the inner structures, even renowned formal-geometric methods, i.e. the Feynman Graphs (Richard Feynman 1918–1988), which are the basis of a mathematical-analytical formulation. Only recently it became possible through new experimental procedures to "observe atomic processes in their own space-time dimensions. One might be eager to see how the visualizations of the non-visual reality of quantum "particles" will further be developed.

Over and above that, the fundamental observer problem followed from the Quantum theories, the insoluble integration of the observer into the observed system on the basis of non-deterministic probability waves functions. This is the first time that maybe the most central problem of all epistemology theories in the natural-scientific modeling occurs, the human observer and the question for his contextual role in relation to the observed phenomenon.

It would be very enticing at this point, but too far reaching, to examine the effective non-compatibility among the Theory of Relativity and the Quantum Theory, especially because here on the one hand the role of the observer has to be considered and on the other hand the problem whether or not there exists a mapping or visualization of the observed phenomenon. Instead of this, we focus on the youngest natural-scientific theory, the theory of chaos or complexity. Whereas the Theory of Relativity and the Quantum Theory were established on the cosmological respectively atomic space and time scale, the theory of complexity challenges the classical imagination of the calculability of the physical models of cosmos.

The change of paradigms from linear to non-linear and therefore potentially unpredictable systems also fundamentally changed the dominance between mathematical analysis and computational science. Non-linear systems are by definition not analytically calculable but need numerical integration to be solved. Thus the computational sciences, the finally image-based multimedia simulations are the only tools to access the non-linear model worlds. All phenomenon and problems of the natural-scientific theories about cosmos mentioned hitherto are now also existing in "human scale" space-time dimensions by the theory of chaos. Experiments like the dropping water tap or the chaotic rattling diesel engine, apart from looking at the local as well as global weather situation, are examples for the deterministic chaos in everyday life. It is already obvious for a longer time that chaotic non-linear systems are the rule and ordered linear systems are the exception to the rule in our cosmos. The paradigms change coming with this knowledge has essential effects on the role which is given to the (human) observer. Complex images and as such numerically exact analysis of non-linear systems are eventually presented, can now only be interpreted by the human observer. I am going into more details following these consequences in the last chapter.

II. Images and Movement

Frequently used but even appropriate as an ordering scheme in relation with New Media is the parallel evolution between technical or scientific knowledge and the corresponding predominant image world. In order to avoid a further strain of such an historical summary, only some of the most important evolution steps of the machine-based image worlds of new media will be mentioned. The idea of an evolution of media is at the same time a definition of new media as a complex system in which the technical and conceptual elements of each step of development are foundations for the following steps.

Let us skip all steps of evolution of Fine Arts concerning sculptural or painted pictures. Our quick tour through the history of machine-based image production begins with the book printing in 1434 (Johannes Gutenberg 1397–1468). Text is printed and paper is the medium. In 1609, the first newspapers were printed and in 1682 printed magazines followed. Certainly, paper remained as the carrier medium but printed pictures completed the texts. Around 1829 the most essential evolution step towards machine-supported picture production came with photography. The medial basis of photographic picture information is since then the chemical carrier.

All text and picture information of print media and photography remains static that means pictures are snapshots of the real event and the content of books or magazines is a condense of the true story behind the texts.

Starting in the middle of the 19th century, the machine-based delivery of information followed the machine-supported picture production, firstly as telegraphy (since 1840) and then simultaneously with the telephone, the telecopy (since 1875). The transmission of audio and visual information was established by the discovery of electro-magnetism. Besides the change of the medium into an electro-magnetic signal, this step also meant the separation of picture and copy. The original remains at its place and the copy is delivered. Nevertheless, the electro-magnetic transport of pictures does not change the fact that the pictures content remains static.

The next step for the development of image worlds was the invention of the film, the machine-moved, chemical pictures which started in 1895. This means about all that now time in form of the illusion of moving film pictures enters the medial images. The chemical space form transforms to a mechanical time form. But the mapping of dynamics is realized until today by acceleration of static single pictures to 25 or 30 "frames per second." By discovery of the electron and the electro-magnetic waves the age of the broadcasting media begins. Since 1897 wireless telegraphs exist and since 1920 the broadcast. Waves as non-material carriers enter at first into the world of electro-magnetic sound pictures. Also the development of the first tapes around 1950 initially only influenced the world of audio. But soon it has been found out that with video tape (since 1951, 1978 respectively) an electro-magnetic medium carrier was found which summarizes reproduction and manipulation under one platform of audio-visual data storage. With regards to the aspect of the dynamics of image worlds it is interesting that the audio information owns a continuous time dependence whereas the picture information is stored and reproduced in single frames despite the analog medium carrier.

One of the big revolutionary steps of this century was the development of television. The audio-visual moving picture quits its chemical carrier and becomes an electronic image. The machine-based production of images, the picture manipulation, the picture storage and the transmission of pictures are integrated in one medial platform. Although it is still true that the images are static and we are only confronted with a simulation of movement, with an illusion of a dynamic movement as the picture worlds are still built single-picture-wise.

The transistor (since 1947/49) and the semiconductor technique (since 1950s) are on the threshold of the present of digital pixel worlds. Today multimedia computer are the medial carrier of the machine-generated simulated model worlds. The entity of the medium is the binary information. Computer integrate all technical and formal possibilities of the historical evolution steps under one platform. Additionally some more essential and new possibilities of machine-generated images follow up. Computers are the first machines which are due to its digital soul space-time simulators.

The multimedia space states of the computer generated simulations are not only mappings or reproductions of our natural environment but they are model worlds of their own. It is for the first time that we have a medium which allows the simulation of dynamic processes, not according to the principle of an illusion of accelerated single pictures but as temporal state spaces or time objects if you will.

Independent from any time scale, computers are time microscopes or time telescopes. Graphical-numerical state spaces are the latest form of machine-calculated image worlds.

III. Images and Interaction

Technologically seen, we presently understand New Media as the electronic media but seen from an information technological point of view we mean the digital media. Today the New Media are consequently: (digital) video or television, (multimedia) computer and the world-wide data matrix of the Internet. The latest step in evolution is the integration of interaction. Machine-calculated image worlds are not only dynamic state spaces but autonomous parallel worlds where the human observer can be in interaction via interfaces. Classical interfaces are the

Fig. 2. Vision from the Knowbotic Interface Project of the INM, which tries to realize computer consciousness. Knowbots are self-learning autonomous agents which develop knowledge of their virtual world (a model of the city of Frankfurt) through understanding speech. Scientists: Gerd Doeben-Henisch and co-workers of the INM, artist: Nicolas Reichelt, © INM-Institute for New Media 1996

haptic interfaces as the keyboard, the mouse pointer or the joysticks of video games. Already under construction are the first sensorial interfaces which will use natural human senses to enable an interaction with simulated model worlds. There is visual interaction via eye tracking, the natural human language or new haptic interfaces as the so-called data glove, a haptic device to use the natural movement of the hand for navigation in virtual space. These developments also try to use traditional human means of communication for the interaction between human user and simulated model world. The major purpose of all these technical developments is the multi-sensorial virtual immersion, the immersion into computer simulated spaces.

All developments described, those of technological kind as well as of information theoretical kind demand for a contemporary definition of "image". How do we then understand the image worlds of New Media nowadays?

The final entity of the new scenarios is the binary information. Objects, states and experiences are processed on electronic devices as binary strings. The new worlds are virtual. The algorithmic access to the binary data enables a manipulation of its contents at every time. Objects and states become variable. Changes of states of virtual worlds and their objects within happen via intrinsic simulation algorithms as well as via reaction to external observer interaction. The character of complex dynamic systems which undergo changes of state either autonomously through back coupling or as reaction to context sensitive inputs from their environment is called viability by the radical constructivism, behavior similar to life. From the triad, virtuality of information storage, variability of image objects and viability of image behavior, an interactive dynamic image system develops, the image as imagination. Primarily from this basic concept, installations of the interactive media art become relevant for the exploration of a new aesthetics of New Media. They integrate one or more human observers in computer calculated virtual scenarios to which computer controlled interfaces, multi-sensorial interfaces are defined. The previously passive observer, external to the object becomes a part of the image world, his behavior influences the virtual scenarios. The simulated worlds react to the observer and the changes within influence by back coupling the observer itself. As a consequence interactive installations replace the traditional understanding of the image as a statically unchangeable object or state with an understanding of the image where the machine generated image becomes a dynamical system of observer manipulated variables. The imagination of the image changes from a window where only a small part of any experience in space and time can be seen, into a door in which the observer enters the world of multi-sensorial experiential fields. These fields describe a contemporal and physical changeable dynamical space of experience. Interactive computer installations realize new worlds of experiences, images as spaces of imagination as the most progressive development step of machine generated illusions of moving images.

IV. Image and Experience

Our imagination of the image has changed due to the latest developments in the New and Interactive Media from a complete figurative or abstract mapping to a

Fig. 3. Screenshot of the Italian House seen from the Wolfs Bridge out of the interactive computer installation of a virtual walk through the Woerlitz park near Dessau, Germany. Artist: Nicolas Reichelt, concept: Michael Klein, © INM-Institute for New Media 1996

multisensorial interactive space of experience. This does not mean that we generate an additional mapping of a reality, however it looks like. On the contrary it deals with objects in timeless and unrestrained spaces which have never been seen before. In virtual computer-simulated model worlds, we are not bound to physical conditions of reality. Cause-effect coupling or spatio-temporal correlations can be newly defined and can be opened for experience. Especially in the world of experiences of the global data matrix phenomenon like the bifurcation of the more parallel telepresence of the individual at different locations are not longer impossible anymore. All these possibilities effect our knowledge of images. The image worlds of our natural environment are entangled with the image worlds of artificial model simulations. The reason for the simplicity of this entanglement is to be found in the fantastic capability of the human brain to adept to any complex environment without regarding its source. It seems to be no question that human brains generate worlds. Whether the human brain transforms sensual experiences of the external environment or whether it is in a state of self-referential dreaming it always works as an information processing or pattern generating system. From this point of view the brain system is built from an external world – our reality or any kind of simulation-, an internal world – an information processing network with algorithmic and memorizing capabilities- and from interfaces between external and internal world, which enable mappings between external and internal realities.

Today these concepts are introduced into the artistic and scientific image worlds. The consequences of this development can rarely be foreseen. To emphasize the possible impact of this development I would like to point out one fundamental problem of the scientific epistemology. All axiomatic natural scientific theories gain their strength from objectivity which is traditionally guarantied with analytical mathematical treatment. But today computer generated visualizations or numerical analysis are often the only chance to handle complex solutions no matter whether we simulate complex theories or whether we evaluate experimental tests. But the human observer is the only instance capable to not only experience but also to interpret images. Until now there is no formal language or objective analytical method to interpret images. For sure the human observer though is a subjective observer. Until today the subjective human observer never was part of any objective theory. Nowadays he became the last resort of truth, what defines the obvious dilemma we are in. Here lies the real challenge of all artistic and scientific understanding of the image worlds and the visions of cosmos of New Media.

V. The Networking Community

Now that we learned about the evolution of images and imagination, what is the actual paradigm of the New Media society?

The most influential concept definitely is the networking matrix of the Internet built of nodes and dendrites connecting these nodes. The Internet should neither be only seen as a technology nor as a simple new medium. It defines the lifestyle of a generation. The matrix is a highly non-linear digital data-space with complex dynamic processes powered by n-players interaction. We should take the challenges and the chances of the virtual space seriously. The INM-Institute for New Media in Frankfurt offers an hot spot to the needs of the artists and scientists dealing with New Media Projects. Artist and scientists are both highly mobile cosmopolitan communities. They live from global exchange of experiences and free floating information. What they need is a hub, a node for the discourse and the know how transfer within the virtual interest group. Besides this there is a permanent demand for temporarily accessible production places, studios and ateliers, equipped with the high tech environment needed to realize New Media projects in science and art. The INM faces this challenge with its own Networking Community concept: the virtual space explorers.

The New Media hub realizes an interface between input and output. It is a dynamic forum, permanently adapting to the demands of fast developing New Media. The creative dispute about virtual space exploration is alive within a multidisciplinary group of scientists, artist, philosophers and application specialist. An active schedule of symposia and exhibitions, workshops and publications, presentations and Internet /web forum allows for an active and interactive method of working. Thus we feel in charge for the "Gestaltung" of the formerly unknown spaces found as virtual extensions or parallel-worlds of our real existence (http://www.inm.de).

8. Art & Science in Historical and Cultural Context

Art and Technology: The Ineluctable Liason

Cynthia Goodman

The liaison between art and technology, both uneasy and inexplicable at times, has produced some of the most remarkable artistic achievements of our century.[1] Although the stereotypical image of an artist is that of a solitary figure in a studio struggling to create works of art in a chosen medium, collaboration has been an essential part of the artistic process from earliest times. Numerous hands, for example, can be discerned in the masterpieces of Paleolithic cave paintings. In Medieval Times, the magnificent glass windows of Gothic cathedrals were designed by one master and then executed by numerous skilled craftsmen. Similarly, in the creation of the Sistine Chapel, one of the hallmarks of Renaissance art, Michelangelo was responsible for the overall vision, yet the actual execution required many assistants.

Utilization of the enormous technological strides of the twentieth century has also necessitated collaborative efforts in order to realize artistic goals. The computer has indisputably revolutionized the artmaking process, challenging many of our preconceptions about how we view, experience and create works of art. Not surprisingly, a reliance on digital technology is inherent within those achievements positioned at the cutting edge of the artistic frontier. The ubiquitous presence of computers in our society, so forthrightly articulated by Nicholas Negroponte in his landmark study *Being Digital* (1995),[2] has elicited a concomitant assumption of widespread expertise in the digital domain. Nevertheless, either due to the scale of a project, the technical complexity or the equipment required, rather than the anticipated independence, technological advancements in art have necessitated an increasingly symbiotic relationship among artists, scientists and engineers in order for artists' concepts to be realized successfully.

The era of computer graphics began in 1959 , when the Calcomp digital plotter, the first commercial drum plotting mechanism became available. Although it might be assumed that artistic applications of the technology would be not far behind, early graphic achievements were sorely constrained by programming

---

[1] The author wishes to acknowledge the omission of many noteworthy collaborations due to spatial considerations. The subject is worthy of a much lengthier study which is currently in progress. For a more extensive discussion of the early developments in art and technology see the author, Digital Visions: Computers and Art, New York City, New York, Abrams, 1987, pp 18–61 and 152–160.

[2] Negroponte, N, (1995) Being Digital. Knopf, New York.

rigors and limited memory. Not surprisingly, the computer graphics of the ensuing decade were dominated by industrial and military applications achieved by scientists, mathematicians and engineers. In fact, it was William A. Fetter, a researcher with the Boeing Company in Renton, Washington, who coined the term "computer graphics" to describe his purely technical cockpit designs. The first time that a computer was used uniquely for artistic purposes is difficult to pinpoint.

Although conducted primarily for engineering purposes, a major breakthrough for both scientists and artists occurred in 1962, when Massachusetts Institute of Technology student Ivan Sutherland completed his now famous doctoral thesis in which he defined his Sketchpad system for interactive computer graphics. The following year, a trade publication, *Computers and Automation*, announced the first contest for computer graphics in which the winners were to be determined on the basis of their aesthetic merit rather than the practicality of their designs. Although a glimmer of hope had been held out to the artistic community, the first and second place winners were entrants from the U.S. Army Ballistic Research Laboratories in Aberdeen, Maryland.

Despite these inauspicious results, one of the great crusaders in the art and technology arena, Billy Klüver, a research scientist in the laser field at Bell Telephone Laboratories in Murray Hill, N.J., was beginning his proselytizing at about the same time. In 1960, Klüver had assisted Swiss sculptor Jean Tinqueley in the construction of *Homage to New York*, a mechanized sculpture that destroyed itself before an intrigued audience in the sculpture garden at the Museum of Modern Art, New York. The success of that event instilled Klüver with a messianic-like zeal to make himself readily available for numerous subsequent art and technology projects. Most publicized of these were his collaborations with New York Pop artist Robert Rauschenberg beginning in 1963. *Oracle*, an environmental sculpture which was the result of three years of teamwork, was particularly significant because both Klüver and Rauschenberg signed the piece, making patently visible the equally important roles of artist and engineer.

Hoping to draw attention to the need for similar artist-engineer collaborations, Rauschenberg and Klüver co-produced an ambitious peformance series called *Nine Evenings* at the 69th Street Armory in New York in October of 1966. Forty engineers from prestigious institutions including IBM and AT&T and ten well known artists including painter Frank Stella and dancer Yvonne Rainier, teamed up to produce the musical, dance and theatrical extravaganzas presented on those memorable evenings. The following year, despite the many technical problems which had plagued their performance series, Klüver and Rauschenberg founded the highly influential organization called Experiments in Art and Technology (EAT). It was their strong conviction that interdisciplinary action between scientists and artists would not only benefit the participants but also all of society. Their belief hit a widespread, sympathetic chord, and by 1970, EAT had over six thousand members nationwide.

As early as 1974, however, Klüver observed that EAT, which had grown "out of the isolation of the artist from the technical world. He did not have any direct access to the engineer, to the scientist. He did not know how to get in touch with them ... [was being] phased out of the operation."[3] According to Klüver:

> People know the language. There are articles in the technical press about engineers who, in one form or another, are working with artists ... The problems associated with just the physical thing of getting artists and engineers together which was the *raison d'etre* of EAT six years ago simply don't exist anymore. The engineering schools are setting up art classes and the art institutes are setting up engineering classes, and so on.
>
> My feeling is that artists' use of technology is here to stay, but that now the real work begins. Some real aesthetic breakthroughs have to happen. If there is to be a viable collaborative situation, art has to be the result ... [4]

Klüver's recognition, that simply collaboration was not enough but also that valid aesthetic products had to be produced in order for the collaboration to be successful, was accurate. Yet, his belief that the increasing availability of computers and relative ease of programming would obviate the necessity for organizations such as EAT to exist was not, and with the waning of EAT's activities, other organizations have arisen to take its place, among them ASCI (Art & Science Collaborations, Inc.) founded in 1988 in New York by Cynthia Pannucci as a network for artists who either use or are inspired by science and technology and the San Francisco-based organization YLEM. Similar organizations have been formed around the world.

Espousing goals similar to those of EAT, The Center for Advanced Visual Studies (CAVS) at the Massachusetts Institute of Technology in Cambridge, Massachusetts, was founded by New Bauhaus Professor Gyorgy Kepes in 1968, to provide a research facility to foster creative interdisciplinary work in art, science and technology. With the vast engineering and scientific resources of the University, a principal beneficiary of industry and government-sponsored research, those affiliated with CAVS have had inordinately rich opportunities for access to advanced technologies. Developing this potential was a mandate of particular importance to Otto Piene, among the early fellows at the Center and subsequently its director from 1974 to 1994. Since its inception, over three hundred Research Fellows, scientists, media designers and engineers, including many at the forefront of the art and technology movement such as Nam June Paik, Tsai Wen-Ying, Stanley VanDerBeek, Scott Fisher, Paul Earls, and Lowry Burgess, have worked at the Center. For Tsai, as an example, the opportunity to work in the late nineteen-sixties, with Harold Edgerton, the eminent electrical engineer and father of stroboscopic photography, was critical to the development of his sound-responsive, stroboscopically-illuminated cybernetic sculptures. Over the twenty year period he was director, Piene's guidance as well as contributions were key to many ambitious international art and technology events beginning with his Sky Art activities in the early seventies, when he collaborated with astrophysicist Walter Lewin

[3] Kranz, S, (1974) Science and Technology in the Arts. Van Nostrand Reinhold, New York, p 55.

[4] Ibid.

in order to calculate the lift forces necessary to float his inflatable sculptures in the sky.

Paul Earls, who has degrees in both art and music has also had an association with the Center for almost thirty years. On numerous occasions, his laser images were projected onto Piene's sculptures, both indoors and outdoors. Earls has always assiduously sought to extend the artistic parameters possible with laser light and electronic music in partnership with other members of the MIT community. Recently, he has worked with Tod Machover, Neil Gerschenfeld and Josh Smith of the Media Lab in addition to programmer Walter Zengerle, on the production of increasingly larger installations of his interactive laser system which controls both music and images in real-time.

CAVS has always considered central to its mission its role as a disseminator of information about art and technology. *Arttransition '90: An International Conference on Art, Science and Technology,* was organized by the Center in the fall of 1990, and the author, then a CAVS Fellow, was conference director. This three-and-a-half day conference focussed on artwork in new media and recent collaborations among artists, scientists and engineers. The participants and attendees including many of the most prominent individuals in the world working collaboratively in these fields, also addressed the growing number of art and technology centers and new media departments in colleges and universities, many of whom had consulted with CAVS in their formation.

Former Fellow and longtime friend of the Center, Nam June Paik, is credited with making the first video art tape in 1965 when he recorded Pope Paul VI visiting New York and showed the now legendary tape that evening at a New York cafe popular with artists. Image manipulation as well as interactivity were important to Paik's artistic concepts from the onset. In *Distorted TV* of 1963, manipulation of the sync pulse transformed the image. Two years later, he created *Magnet TV,* for an exhibition at the New School for Social Research in New York. Placing a large magnet on the top of the television set in this work and moving it disrupted the electronic signal and created both distorted images as well as abstract patterns.

In 1970, Paik developed his exploratory image manipulation further by collaborating with engineer Shuya Abe to produce the *Paik/Abe Synthesizer*, the first broadcast quality synthesizer. This achievement advanced Paik's exploration of video imagery from the level of artistic exploration to a functional broadcast tool. The *Paik/Abe* was also the first of many subsequent collaborative efforts, and Paik has always both quickly and generously acknowledged the contributions of assistants.[5]

---

[5] Recent large scale laser works have required the assistance of a fairly large team of skilled engineers in addition to electricians and video technicians. For *Laser Experiment No. 5* recently on view in the InfoART Pavilion at the '95 Kwangju Biennale in Korea as well as subsequent viewings in Europe, Paik collaborated with laser artists Horst Baumann and Norman Ballard. He also acknowledges the critical role of the 3-D glasses invented by Jeffrey Cone through which the installation is viewed. For additional documentation of this installation see Goodman C (1996) InfoART. Rutt Video Interactive, New York (CD-ROM).

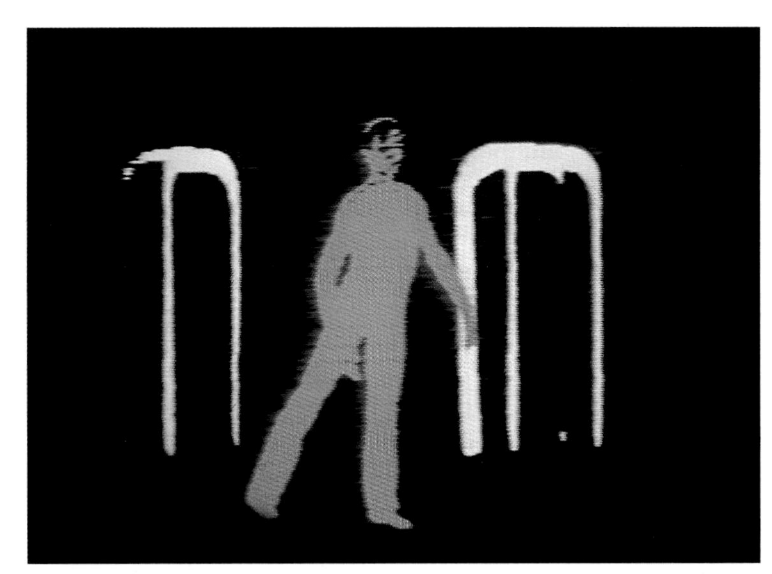

Fig. 1. Dance Nine. © Doris Chase

Video art was very much a product of the nineteen-sixties. In addition to Paik and Abe, other artists and engineers interested in pushing the limits of video to create new effects invented a number of now historic devices. These somewhat idiosyncratic, hand-built machines were responsible for seminal work in the field and enormously expanded the potential of this new medium as well as opened the way for personal expression. Among these historic devices were the *Rutt/Etra Scan Processor,* Stephen Beck's *Direct Video Synthesizer* and Dan Sandin's *Image Processor*. The necessity of first developing the tools was underscored in a statement by Bill Etra who collaborated with his wife Laurie Etra and Steven Rutt to produce the Rutt/Etra Scan Processor: "For my own work, I never produced more than ten minutes a year I ever showed, and that's an awful little, but it's an awful lot if you think that most of the machines they were shown on had to be built before the tapes could be made."[6]

Rutt and Etra were inspired to build the *Rutt/Etra Processor* after seeing the Paik/Abe at Channel 13. Rutt, who had grown up in a household in which both electronics and art were present, was both responsive and receptive to helping artists articulate their visions electronically. Bill Etra, a video artist, Rutt remembers acted as, "the interface to those in the art community who were using the synthesizer."[7] Woody Vasulka has described Rutt's factory as "a special case, something we all wanted to have exist, something where the artist would participate directly in toolmaking and which would facilitate the cultural continuity of invention we know and treasure in photography, film, and video."[8]

[6] Etra, W, Interview with Darcy Gerbarg, New York, quoted in Goodman, C., Digital Visions, p 168.

[7] Rutt, S, Interview with Cynthia Goodman, February 23, 1997.

[8] Dunn, D, (ed) (1992) Pioneers of electronic art. The Vasulkas, Inc., Sante Fe, New Mexico, p 136.

Vasulka has also stated that the "instrument called *Rutt/Etra*..was a very influential one...Almost everybody I respect in video has used it at least once ..."[9] Among those was sculptor, dancer and video artist Doris Chase, who produced some of the most fruitful artistic compositions with the *Rutt/Etra* assisted by Steve Rutt. In the video tape *Dance Nine*, for example, the movements of dancer Gus Solomons are accompanied by those of an abstracted video image of one of Chase's streamlined arc-shaped sculptures. To create Solomons's electronic dance partner, Chase cut out an arch like one of her sculptures and then placed it on a black background. This silhouette was shot by a camera attached to a Rutt-Etra synthesizer controlled by Rutt, who in response to "... verbal cues from Chase, multiplied, bent, stretched, collapsed, and otherwise manipulated the arch into a variety of moving patterns and shapes."[10] Subsequently, Chase incorporated segments of this tape not only into *Dance Nine*, but also in live performances as rear projections behind dancers with her sculptures on which the abstract figure was based as props on stage.

Whereas the *Paik/Abe* and the *Rutt /Etra* both represented the joint efforts of an artist and an engineer, the *Sandin Image Processor* or *IP*, which was built by Dan Sandin functioned in both capacities. With undergraduate and graduate degees in physics, Sandin, whose father was a photographer, realized while working on his masters degree, that what really interested him was "working in images." Athough he did the construction by himself, Sandin credits video artist Phil Morton, who was the first to take advantage of his offer to give away instructions about how to construct the machine to anyone who asked him for them, with playing a major role in the ultimate production. Morton's request initiated a year of Friday afternoon sessions during which he and Sandin prepared the complete documentation with schematics which made it practical to do so.[11]

Woody and Steina Vasulka are synonymous with the emergence of video as an art form. Woody remembers the impact of video upon him when he was first seduced away from film by this raw, malleable and began to use it in the late nineteen-sixties: "When I first saw video feedback, I knew I had seen the cave fire. ...a perpetuation of some kind of energy." [12] Steina, who discovered video with a similar rapture, came to the medium from a background as a concert violinist with the Icelandic Symphony. Together in 1971, they founded the highly influential Kitchen in New York City, a center where video and performance art were shown and which was catalytic for many early efforts in both media. Collaboratively and independently, the two have explored the relationship of electronic audio and video. Their experiments with the various video systems available as well as with the Digital Image Articulator which Woody Vasulka designed with Jeffrey Schier are among the most powerful and memorable in the medium.

---

[9] Ibid.

[10] Falling, P, (1991) Doris Chase artist in motion: from painting and sculpture to video art, University of Washington Press, Seattle and London, p 81.

[11] Sandin, D, Interview with Cynthia Goodman, February 22, 1997.

[12] Dunn, D, p 83.

Unlike many others in the field who have moved into other media, the Vasulkas have remained steadfast in their exploration of video while incorporating digital image synthesis and interactive multimedia into their basic video vocabulary and its display in both single channel works and interactive multimedia installations.

Today, the use of digital imaging techniques by video artists is commonplace. For a number of years, artist Paul Garrin has assisted Paik in the realization of the computerized video effects which are incorporated in Nam June Paik's sculptures and video walls and are a signature of his style. Recognized by Paik for his mastery with digital video techniques, Paik unabashedly refers to the extraordinarily talented Garrin as "the genius behind my work."[13]

In his own interactive video installations, Paul Garrin utilizes contemporary technology for the exploration of trenchant political and sociological themes. Two recent works, *White Devil* and *Border Patrol (1995)*, are reliant on a tracking device developed by fellow artist David Rokeby in order to record and process visitors' movements. Rokeby's enlightening comments about their intense collaboration shed light on many such team efforts. According to Rokeby:

> On one level, working with Paul is very similar to working with myself. I am often my own engineer, drawn by that other side in my crazy desire to make something apparently impossible happen. ... One particularly interesting thing about collaborating is that I find ways of using technology that I have developed in ways I never would have imagined. New algorithms, borne out of urgent necessity have provided real breakthroughs. Things I might have assumed were impossible turned out to be very doable. Someone (usually the artist) just has to want it enough, and the technician has to be able to step out of their particular mindset for a moment and imagine a totally different way of thinking about the problem. Many artist/technician projects fail because of lack of communication and an accompanying lack of respect. Working with Paul has been successful because we both have a feel for technical things, and we are both artists.[14]

Rokeby's intrigue with the real-time processing of images and sound, has been a tenet central to his inventive artistic explorations since the early nineteen-eighties, when he began developing his *Very Nervous System*, an increasingly complex and sophisticated tracking apparatus. Rokeby's most recent installation *Watch (1995)*, incorporates two surveillance cameras: one within the exhibition space and one in an adjacent area.Visitors quickly become aware by observing the images projected on the walls that not only are their movements being recorded but also those of unidentifiable others. Like Garrin, Rokeby uses a technological vocabulary for wrestling with provocative social commentary and human response. Moreoever, the very nature of his interaction with the technology is a critical compositional component. In contrast to the experience of helping Garrin implement his artistic goals, in his own work, he explains, "the installation and the technology are much more closely inter-woven. One of my big questions as an

---

[13] Paik, N J, Interview with Cynthia Goodman, May 10, 1995.

[14] Rokeby, D, Correspondence with Cynthia Goodman, February 19, 1997.

artist has to do with how these technologies change and guide us, for good and for bad." Rokeby readily admits to being intensely aware "of the significance of every hardware and software decision that I am making. Each of these technical decisions tangibly changes the final artwork, whether I intend it or not. With Paul things are much more practical ... make it work!" [15]

In the past decade, the "real aesthetic breakthroughs" envisioned by Klüver have become manifest in a group of works distinguished by an ineluctable and magical quality which seems to arise when a seamless interface has been created between artwork and technology, and the resultant composition is informed equally by a profound knowledge and respect of both science and art. These Interactive Computer Installations which are fairly elaborate and challenging both visually and technically utilize virtual reality (VR) techniques. The revolutionary new art forms created with VR are contributing to the transformative impact which information technology is having on our entire lives, making not only "the basis for a new culture of simulation,"as Sherry Turkle describes the phenomenon in her recent book, *Life on the Screen,* but also a "fundamental reconsideration of human identity." [16]

Artist and computer scientist Myron Krueger is one of the pioneers in virtual reality, and his writings have contributed significantly to the public's awareness of the field. His 1974 doctoral thesis, "Computer-Controlled Responsive Environments," is a seminal work on human-machine interaction as an art form. Although the belief is widely held that early experimentation and development of virtual reality is the exclusive domain of the scientific community, Krueger convincingly postulates in a later essay called "The Artististic Origins of Virtual Reality," how the arts community has also been an active contributor to developments in the VR field since the early sixties. According to Dr. Krueger, not the technical community but "...artists were the first to recognize that virtual reality was of fundamental importance. Rather than publishing reports to the scientific elite, they communicated it immediately to the public through their work. " [17] Among the other pioneers whom he mentions are filmmaker Morton Helig whose Sensorama theatre, built in the early sixties provided a fully immersive experience including stereo film, stereo sound accompanied by movement and olfactory stimuli and Dr. A. Michael Noll, who pioneered three-dimensional movies seen through a stereo viewing apparatus. Krueger also discusses *Glowflow*, an artwork which he and Dan Sandin created in 1969 when both were graduate students at the University of Wisconsin, as a seminal work in computer-controlled responsive environments.[18] Lastly, Krueger notes as significant the fact that the data glove was invented in 1978 by three leaders in the computer art community, Dan Sandin, Tom DeFanti and Gary Sayers, under a grant from the National Endowment for the Arts.

---

[15] Ibid.

[16] Turkle, S, (1995) Life on the screen: identity in the Age of the internet, Simon and Schuster, New York.

[17] Krueger, M, (1993) The artistic origins of virtual reality, Typescript, p 2.

[18] Krueger, M, (1983) Artificial reality, Addison-Wesley, Reading, Massachusetts, pp 12–17.

Another seminal figure in in the development of artistic applications of virtual reality is University of Paris 8 Professor Edmond Couchot. His hallmark work, *The Feather,* was created between 1988–1990 with the help of flight simulation specialists from SOGITEC in France. For a more recent work of 1990, *I Sow to the Four Winds,* he collaborated with Michel Bret (who wrote the software) and Marie-Helene Tramus, fellow researchers and professors at the University. Both interactions which are graphically displayed as large-scale projections in a darkened room are activated when the participant blows into a small microphone mounted on a panel of Plexiglas. The speed at which *The Feather* is propelled into flight is determined by the strength of the person's breath. Whereas, in the case of the latter work, blowing into the microphone causes clumps of dried dandelion seeds to be detached until the stem is barren. These deceptively simple yet elegant works exemplify how high-tech tools can inspire poetic reverie with equal effectiveness as more dramatic effects.

A select number of artists have been given research positions in advanced centers of computer imaging around the world. Among these fortunate artists are Christa Sommerer and Laurent Mignonneau, who for the past several years have been invited researchers in the MIC Lab at the ATR Advanced Telecommunication Research Laboratories in Kyoto, Japan. Mignonneau and Sommerer endow their collaborative interactive installations with a naturalistic splendor rarely equalled in the multimedia field. Both artists and both technologically astute, in their case sophisticated graphic tools are in the hands of users whose artistic capabilities are commensurate with the power of their medium. Their work is strongly influenced by Sommerer's studies in botany and biology at the Academy of Fine Arts, Vienna, and her post-graduate work at the Institute for New Media in Frankfurt, Germany, where she met Laurent Mignonneau with whom she has collaborated since that time. Prior to his post-graduate work at the Institute for New Media, Mignonneau studied fine art, video and computer art at the Academy of Fine Arts in Angouleme, France.

Most developments in the computer art world keep apace with developments in the industry. Yet Mignonneau and Sommerer have been able to push the limits of the existing technology so that their work not only delights the general public but also intrigues sophisticated computer professionals. In so doing, they are outstanding proof of Krueger's contention that artists have led the way in the VR field. A recent piece called *Trans Plant* of 1995 resulted in the patent of a new 3-D video key. Their work is the product of a lot of give and take both technically and artistically. As Sommerer has described their collaborative process:

> Basically we develop the concept for a new work together and then look for technical solutions to its realization. Laurent has great knowledge in electronics and programming which helps us find technical solutions in the search for interfaces. When it comes to programming, Laurent writes the main structure of the computer code; whereas I work on the design of the programs and modify the code in terms of shapes, colors and general look. For the interface design, Laurent works on it alone. Once the technical problems are solved, we usually readjust the work and

collaborate on the interactive part: we test the system and try to see how people interact with it. After that test period, we generally do some readjustments.[19]

Because of the complexity of their compositions, Sommerer stresses that is important for them to know in advance "what technology is available, what can be assembled, and in case it is not available, how we can invent something new."[20]

Monika Fleischmann, Wolfgang Strauss and Christian Bohn, also fortunately have access to the resources of the German National Research Center for Information Technology (GMD) in Sankt Augustin, Germany, with which they are all affiliated. Fleischmann, who is artistic director of Visualization and Media Systems and head of computer art activities at the GMD, has a multidisciplinary background in fashion design, art, drama and computer graphics. Strauss, a guest researcher at the GMD, is an architect and media artist in the fields of virtual architecture and interface design; and Christian Bohn is a computer scientist responsible for scientific visualization, neural networks and virtual reality at GMD. He is also conducting research on radiosity rendering. According to Fleischmann, her primary contribution is the development of the concept or story, although she discusses it with her partners. Wolfgang "brings a lot of his architectural background to the installation itself and Christian the scientific visualization as well as the capability of the hardware for special input."[21]

Their varied skills provide a rich range of resources which have fortified them when tackling difficult themes such as the body as interface to the virtual world. Their memorable interactive installations including *Rigid Waves* and *Narcissus*, both of 1993, are artistically and dramatically staged and constructed as well as visualized using the most sophisticated computer graphics techniques. They see their role as developers of responsive environments with "challenging Human-Computer Interfaces" as a "means of artistic and scientific articulation supporting the process of imagination."[22]

Unlike other programs where artists and engineers or scientists are brought together for the purpose of collaboration, at the Electronic Visualization Laboratory at the University of Illinois at Chicago (EVL), headed by Dan Sandin and Tom DeFanti, students are trained equally in both fields. As a consequence, a breed of artist/engineer is being developed with equal proficiency in artistic conception and programmatic execution. The visionary accomplishments of the EVL are powerfully illustrated by the CAVE (TM) a multi-person, room-sized, high-resolution, three-dimensional video and sound environment developed by Sandin, Carolina Cruz-Neira and DeFanti, with other staff members as well as students. In its current state, four projectors throw full-color computer-generated imagery onto three walls and the floor. All the perspectives in the CAVE have been calcu-

---

[19] Sommerer, C, Correspondence with Cynthia Goodman, February 9, 1997.

[20] Ibid.

[21] Fleischmann, M, Correspondence with Cynthia Goodman, February 17, 1997.

[22] Fleischmann, M, Imagination systems: interface design for science, art and education, Visualization and Media Systems Design German National Research Center for Information Technology, Sankt Augustin, Germany, p.1.

Fig. 2. CAVE. © Board of Trustees, The University of Illinois

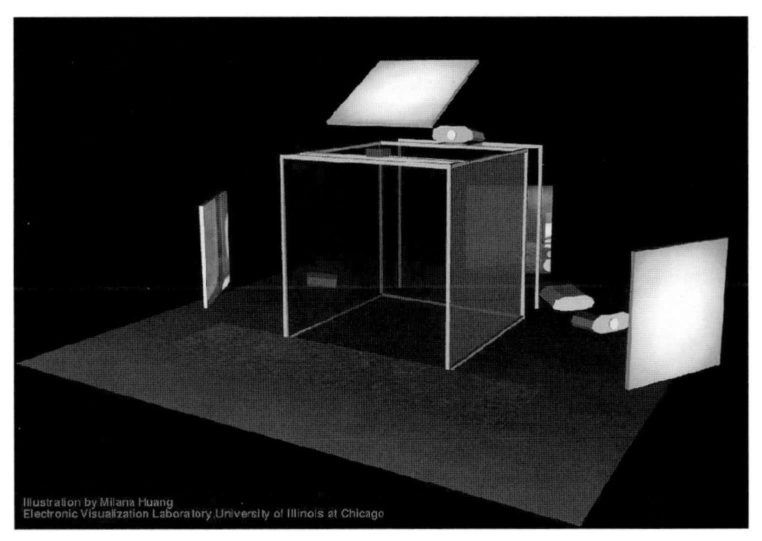

Fig. 3. CAVE. © Electronic Visualization Laboratory, University of Illinois at Chicago, Illustration Milana Huang

lated from the point of view of the viewer who wears active stereo glasses which alternately block the right and left eye. The participant in the CAVE experiences an unprecedented sensation of immersion in a room-sized environment while navigating a wand which transports him or her from one part of a scene as well as from one of the many visually compelling, fanciful worlds which have been created for this system to another.

Dan Sandin has also commented on the astonishing similarity of the situation in the art and technology worlds thirty years ago and now, when it is still hard for individuals to own and maintain the requisite equipment to make cutting edge work; so pooling and sharing resources was and is natural. Despite the challenges, Sandin enjoys maintaining his status as a pioneer. In his explanation: "On the expanding horizon, you're still pushing the technology, the economics and the poetics (by which I mean the goals and how you express them in this new medium)."[23] Then and now, Sandin was among those who were not only defining the artistic horizons but also inventing the devices which enable users to express goals in a new vocabulary. What he feels is particularly advantageous about working in a "virgin medium" despite the obstacles, is the ability to make "broad early statements in a wide open vista neither informed nor constrained by the history of the medium ... and test the technological capabilities of the medium and see what can be done with it which couldn't be done previously."[24]

Director since 1991 of the Institute for Visual Media at the ZKM Center for Art and Media in Karlsruhe, Germany, one of the leading facilities in the world with a mandate for the realization of art and technology projects, Jeffrey Shaw has a long history of working with state of the art tools in the realization of his art works. In the case of *The Legible City (1990)*, which he created with Dirk Groeneveld, his collaborator on several earlier works, Shaw's role was similar to that of the producer of a film. This complex piece was produced in several cities not unlike the basic concept of a second version of the piece in which the user who is seated on a high-tech bicycle can choose to navigate the streets of one of three cities – Amsterdam, Manhattan or Karlsruhe. Shaw was living in Amsterdam during the construction of the piece; whereas, the graphics were rendered in Los Angeles by Jerry Wells and Karl Sims at OptiMystic, a computer graphics facility.

Once the bicyclist has selected a site, he or she "rides" through a computer graphic simulation of that city based on actual ground plans maps. In Shaw's City, the actual buildings and other distinguishing characteristics have been replaced by computer-generated 3-D forms which assume the shapes of letters and texts along the sides of city streets. In recent years, not only has Shaw continued his exploration of state of the art imaging techniques in his own works but his guidance has also been instrumental to the many artists who have come to the Center as the recipients of fellowships in order to produce their own works of art.

With the "recreation of a paleolithic sanctuary for contemporary art audiences,"[25] as his goal, University of Cincinnati Professor Benjamin Britton's virtual reality construction of the caves of Lascaux, France, which are now closed to the public, required an interdisciplinary team of about twenty-five members in order to realize. This large group has included not only computer programmers and

---

[23] Sandin, D, Interview with Cynthia Goodman, February 22, 1997.

[24] Ibid.

[25] Britton, B, Interview with Cynthia Goodman, June 3, 1995.

engineers but also prominent archaelogists. Wearing VR goggles and navigating via a spaceball tracking mechanism, visitors to this three-dimensional computer-generated cave have the sense of total immersion on their passage through its winding corridors decorated with masterpieces of paleolithic painting.

Fig. 4. LASCAUX virtual reality art installation: Hall of the Bulls. © 1997, Benjamin Britton

Begun in 1990, a significant amount of the work on *LASCAUX* was done when only rather primitive tools were available for creating VR, and thus much of the software had to be authored by Britton and his co-workers including his brother David Britton, a computer programmer. The complexity of the project, and the consequent necessity for intricate collaboration between the artists and scientists who were involved, gave Britton not only increased respect for his scientific counterparts but also an increased awareness of the diametrically different attitudes inherent in science and art: "Whereas scientists are often blinded by their search for truths, artists are often blinded by their passions. Nevertheless, both are inspired by a common curiosity and desire to convey their perceptions about human life to the public."[26]

Relentless in his quest to improve upon his portrayal of the cave, Britton keeps abreast of all potentially applicable technological advances and incorporates them

[26] Britton, B, Interview with Cynthia Goodman, March 2, 1997.

as quickly as it is feasible to do so. As a consequence, *LASCAUX* is still undergoing refinements in 1997, although various versions of it have been installed in numerous locations around the world both in museum environments such as the InfoART Pavilion at the '95 Kwangju Biennale, co-organized by Nam June Paik and the author, at Walt Disney's EPCOT as well as at various international conferences including SIGGRAPH '96 in New Orleans. A video artist and a professor of electronic art, Britton and his devoted team with varied computer graphics and modelling skills, have achieved several virtual reality innovations including the "first 'dissolve' from virtual reality to a video motion sequence."[27]

Another trained artist who has made a major contribution to the virtual reality world is Scott Fisher, President of Telepresence Research, in Portola Valley, California, a company committed to finding practical applications for virtual reality. After an undergraduate training in fine arts, Fisher went to M.I.T., where he was affiliated with CAVS and the Architecture Machine Group directed by Nicholas Negroponte and now called the Media Lab. He credits a desire to develop the stereo images which he was making there with causing the turn in his career from fine arts to the creation of hybrid art forms that originate at the convergence of science and art. Recently Fisher, who directed the Ames Virtual Environment Workstation Project at NASA from 1985–90, has been collaborating with artist Perry Hoberman on projects such as *Chimerium* (1995), a whimsical world of metamorphosizing denizens controlled by users via VRML. According to Fisher:

> The design and implementation of interactive art experiences in new media such as VR and Telepresence is best developed as a team-based effort similar to that found in the movie industry – this is faster evolving a new kind of working relationship among the traditional roles of artist, scientist, engineer and programmer. And often, each has to play multiple roles in collaboration on key design decisions that in turn drive development of unique custom hardware configurations.
>
> In our recent projects ... it is the technical implementation phase of our projects that usually tends to polarize our respective inputs towards a split between techical direction and creative direction – artists/engineers of necessity become managers and producers while trying to maintain focus on the importance of the final interactive experience. There's great satisfaction in exploring these new possibilities but also tremendous frustration when more traditional views insist on crediting only the producer or the artist rather than the collective team effort. For me, it's the interchange and synthesis of these disparate viewpoints and vocabularies that really make the collaboration work.[28]

At the close of the twentieth century when cybertravel often successfully replaces conventional modes of travel, and virtual reality is embellishing our everyday reality, a televirtual society is emerging with such blinding speed that imminent obsolescence is the only given for any emergent technology. Many artists have espoused the belief that a decreased reliance on scientists and engineers gives

[27] Britton, B, Unpublished manuscript, February 26, 1997.
[28] Fisher, S, Correspondence with Cynthia Goodman, February 24, 1997.

them the potential for enhanced artistic expression. In my evaluation, the converse is actually true, and many of the most provocative accomplishments of our time have resulted from the confluence of the two worlds. These achievements are all the more remarkable, when one considers that despite the growing recognition and acceptance not only of the very validity of such hybrid art forms but also a belief in their very necessity, there are still many obstacles – both scientific and artistic – to overcome. Artist David Rokeby has revealingly expressed the tenuous nature of the successes produced by this liaison. In his words, regardless of "... all the talk of art/technology convergence, the space between the cultural and the technological, the artists and the engineer, is not an easy space to navigate."[29] Yet this convergence is essential for artistic cyberexploration to proceed.

[29] Rokeby, D, Correspondence with Cynthia Goodman, February 19, 1997.

Seeing at a Distance – Towards an Archaeology of the "Small Screen"

Erkki Huhtamo

Living our lives surrounded by screens has, according to some, become our second nature. Can we imagine a day during which we would not at least fleetingly glance at a television set or an animated billboard, work in front of a computer display, play a video game, use a teller machine or a portable personal organizer, or perhaps go to the movies? Screens are becoming a new artificial environment, an extension of the physical urban space and a surrogate reality.[1] The very proliferation of screens has the effect of making them transparent; we use them but we hardly pay attention to them (unless they unexpectedly go blank). They seem self-evident, as if they had always existed. For the younger generations, especially for kids and adolescents who have grown up with personal computers and video games, this may be literally true.

We assign screens many functions. Some of them entertain us, others display information for practical professional purposes from business to healthcare and surveillance. Sometimes we are invited to "break through the screen" and plunge into virtual worlds that purportedly lie beyond it. In other cases the data is sprinkled right at our faces, or just marches past our eyes. Some screens are used interactively, engaging us in a conversation. Others provide us with events and data meant to be consumed more or less passively. The screen technologies vary. There are screens of many sizes and shapes. Some of them are overwhelmingly large, others are small and still constantly shrinking. There are also "screenless screens", displays that abandon the idea of showing images on a flat material surface and replace it with mirage-like three dimensional representations. Whether it makes sense to refer e.g. to a holographic display as a screen is open to argument.

As this variety already implies, the screen, as any element of media culture, does not have an immutable identity. Screens are cultural artifacts that have their own histories, conditioned by cultural, social, technological and ideological factors. From such a perspective the current omnipresence of screens is far from self-evident. The situation has been caused by the gradual saturation of culture by media technologies, many of which emphasize the visual. The screen provides a

---

[1] Callas P (1990) Some liminal aspects of the technology trade. Video screens versus horizon in Tokyo and New York. Mediamatic 5 (3): 107–115

mediated experience; it is not a goal in and for itself. Its purpose is to represent and mediate something else – information, communication, figures, faces and virtual realms. In other words, the screen is an interface device, a mediator by means of which we get in touch with other (ir)realities, either in real time or in other temporal modes.

Considering the centrality of the screen in the culture of our century, it is surprising that no history of the screen exists. There is no comprehensive account about how the 20th century ideas about it came into existence and have developed and influenced each other. The book that gets perhaps closest is Siegfried Zielinski's *Audiovisionen* (1989), a meticulous study about the interplay between cinema and television in the 20th century culture.[2] Film theoretical studies have dealt with the screen as part of the cinematic apparatus.[3] Film historical studies have pointed out some of the links that connect the "big screen" with earlier practices, such as the shadow show and magic lantern projections.[4] The origins of the "small screen" (the television, the radar or the computer display) are more obscure, although some light has been shed on this issue by Lev Manovich. The main focus of Manovich's groundbreaking work, however, is on the 20th century developments of the electronic screen.[5]

It would be tempting to connect the appearance of the small screen with the coming of modernity itself. Indeed, the image on the small screen *is* normally an electronic image, made possible by the invention of the cathode ray tube around the turn of the century. Television and the computer display would thus be genuine 20th century inventions, with no predecessors. This article shows that such an interpretation would be limited and misleading. Looking at the history of the small screen merely in terms of technological development (advances in electronics, etc.) lapses into a deterministic view which sees technology as the prime mover of cultural development. Although the development of the cathode ray tube unquestionably played an important role in the formation of the television set, determining some of its features and influencing aspects of the viewing situation, such an explanation would be limited. It bypasses the fact that the invention and the development of the cathode ray tube happened in a cultural situation in which the idea of the small screen had already been anticipated.

The electronic small screen had concrete predecessors, but, perhaps even more importantly, there were discursive manifestations that had elaborated on the idea. It could be claimed that discursive screens, the ones that were switched on only in imagination or on paper, influenced the development as much as the materialized ones. The artifacts and the discourses that anticipated the appearance of

---

[2] Zielinski S (1989) Audiovisionen. Kino und Fernsehen als Zwischenspiel in der Geschichte, Rowohlt, Reinbek bei Hamburg

[3] See de Lauretis T, Heath S (eds) (1980) The Cinematic Apparatus. Macmillan, London

[4] See Belton J (1992) Widescreen cinema, Harvard University Press, Cambridge, MA

[5] Manovich L (1993) The engineering of vision from constructivism to virtual reality, Doctoral Thesis, Rochester, N.Y.: University of Rochester, College of Arts and Science (unprinted). This manuscript is being revised for publication by the University of Texas Press under the title, The Engineering of Vision: from Constructivism to the Computer.

the small screen have been largely ignored, perhaps as part of the general *damnatio memoriae* of the Victorian technological culture. The idea of the modern, as epitomized by the declamation of the futurists, emphasized the necessity of a break with the decadent and obsolete cultural forms of the past. The brave new media society was equated with the dawning of a new century. The emergence of the electronic small screen seemed perfectly in harmony with this development. Victorian cultural forms were doomed into oblivion. Highly influencial 19th century "media machines", like the magic lantern and the stereoscope, were largely overlooked by early 20th century critics. As a consequence, phenomena like the cinema seemed to appear from out of the blue. Now at the end of the 20th century we should be in a position to re-estimate the technological achievements of the Victorian era. Yet, as the following historical "excavation" of the small screen shows, we should not be content with this. We should travel further back in time, to the 16th and the 17th centuries at the very least.

The Task of Media Archaeology

In what follows I will make some notes about the formative development of the small screen from a media-archaeological point of view.[6] I will investigate some of the material and discursive factors that anticipated and influenced the coming of the electronic screen in the early decades of the 20th century. By media archaeology I mean a research approach which emphasizes the multi-layered, "polyphonic" nature of cultural processes. Technological innovations, such as the small screen, are always embedded in complex conglomerations of discourses, expressing the aims, policies, hopes, fears and desires of an era. These discourses give artifacts their signification and position(s) in the fabric of culture. The formal structure and look of the artifacts themselves also bears evidence of the discursive production surrounding them. Petrified as fragmentary traces of the past (images, texts, clothes, everyday objects) these discourses constitute the primary material for the media-archaeologist.

Media archaeology leads away from the traditional, artifact-centered approach to technology. Technology is not conceived in terms of a linear, progressing series of gadgets that make their predecessors obsolete. Recording engineering feats and breakthroughs in hardware design is secondary. The important thing is to understand the role of technology (in this case the technology for showing and/or transmitting images by means of specifically designed surfaces) as one of the ingredients in cultural processes. These complex processes consist of highly divers elements. While some of these elements seem to have originated just recently, most of them prove out to be conventional, repeated countless times in different guises. The media archaeologist sees the past as an inexhaustible storehouse of cultural formulas which have been activated again and again in the course of history. These formulas, or *topoi*, represent the iterative, conventional aspect of

[6] For a more extensive treatment, see my "From Kaleidoscomaniac to Cybernerd. Notes toward an archeology of media". In: Druckrey T (eds) Electronic culture. Technology and visual representation, Aperture, New York, pp 296–303, 425–427

culture, often lurking behind the innovative and the unique.[7] Things that seem at first sight unique breakthroughs, may prove to be just newly packaged manifestations of old and widely distributed *topoi*.

To sum up, a media archaeologist does two things: he traces the cultural "life" of the *topoi*; at the same time he tries to understand how and why these *topoi* have been activated again and again in certain times and places. Accordingly there is a double focus, one temporal and the other spatial: to trace the appearances of formulas across time and to (re)place them in specific contexts. Because of this media archaeology is concerned with the present as much as with the past. It sheds light on contemporary uses of technology while it also clarifies earlier cultural formations. A case in point, virtual reality was marketed as a sensational novelty in the early 1990's; it was not just a technology, it was a surrogate reality, a new mode of being, a new ontology. Media archaeology can show that the discourses on virtual reality were far from unique; similar hopes had been attached to earlier technologies, as demonstrated by the discourses around immersive cinema spectacles in the 1950's or those triggered by stereoscopic photography during the Victorian era.[8] To understand what is really new and groundbreaking one has to consider its reverse as well.

Big Screens and Peep Holes

If we review the screen practices that preceded the appearance of the television and its "picture window" in the early 20th century, the endeavour of enlarging tiny images by projecting them on a canvas or a wall with a dedicated apparatus seems to dominate. Above all, there was the culture of the magic lantern projections, which began in the 17th century and reached its greatest blossom during the second half of the 19th century.[9] Lanterns were everywhere. In many countries there were educational institutions that organized high quality lantern presentations for amusement and profit. There were also professional showmen who travelled from city to city, from village to village showing lantern programs – travelogues, popular scientific topics, sentimental moral stories, fairy tales, even "risky" topics. The audience sat in front of the screen. It devoured projected images, enlivened by the narration of the showman and perhaps by the tunes played by a musician.

When the Lumière brothers and other film pioneers began to project films in the mid 1890's on a stretched-out canvas in a cafè or some other location, they

[7] Topoi (singular: topos) can be considered as formulas, ranging from stylistic to allegorical, that make up the "building blocks" of cultural traditions; they are activated and de-activated in turn; new *topoi* are created along the way and old ones vanish to show up again in other circumstances. See my "From Kaleidoscomaniac to Cybernerd. Notes toward an archeology of media".

[8] See my studies "Encapsulated bodies in motion. Simulators and the quest for total immersion" (1995) In: Penny S (eds) Critical issues in electronic media, State University of New York Press, Albany, pp 159–186 and "Armchair traveller on the Ford of Jordan. The home, the stereoscope and the virtual voyager" (1995) Mediamatic (Amsterdam), "Home" issue, Vol. 8 (2–3): 13–23.

[9] See Robinson D (1993) The lantern image. Iconography of the magic lantern 1420–1880, Nutney, East Sussex: The Magic Lantern Society of Great Britain. See also Mannoni L (1994) Le grand art de la lumière et de l'ombre. Archéologie du cinéma, Nathan, Paris

could rely on this tradition. Indeed, originally the film projector meant just the device which allowed the celluloid film to be run in front of the lens of an ordinary magic lantern; in a way it was just an accessory. Many early film people had a background in the magic (or "optical") lantern business, either as professional lanternists, instrument makers or producers of lantern slides.[10] There were also other related experiences: rolling panoramas, dioramas, shadow shows – in spite of the differences, in all of them the audience was seated in front of a canvas on which the spectacle unfolded. There was a general tendency towards larger screens during the 19th century.[11] This can be explained by a number of factors, both technical and social. New powerful light sources, particularly the oxyhydrogen or the limelight, were introduced which made projections in large halls viable.[12] With candles and even with the improved Quinquet oil lamps this had not been possible.[13] Simultaneously the social development during the 19th century led to a need for spectacles which could accomodate larger audiences, both among the bourgeoisie and the working class.

This growth reached its symbolic apogee with the giant screen installed by Louis Lumière at the giant ballroom of the Paris World Fair in 1900. The photographic lantern slides *and* films (!)projected on it were seen by millions. Yet although the screens for projecting images were mostly associated with the public sphere, they entered the privacy of the home as well. Small, ornamental toy lanterns became popular after the 1840's, particularly in France and Germany. The protagonist of Marcel Proust's novel *Du Coté de Chez Swann* recalled a domestic lantern projection, experienced in the children's room: "[E]lle substituait à l'opacité des murs d'impalpables irisations, de surnaturelles apparitions multicolores, où des légendes étaient dépeintes comme dans un vitrail vacillant et momentané".[14] Domestic lanterns produced a faint, rather small image, much like the professional lanterns of earlier centuries. In a way the size of the audience was limited by light. Curiously, this necessary intimacy anticipated some aspects of the television spectatorship.[15]

To conclude, however, that the consumption of images in the 19th century

---

[10] See e.g. Gray F (ed) (1996) The Hove Pioneers and the Arrival of Cinema, University of Brighton, Brighton

[11] However, the screen size in shadow shows remained relatively small, limited by the nature of the spectacle: the figures were animated by human beings.

[12] See Guerin P (1995) Du soleil au xenon. Les techniques d'eclairage à travers deux siècles de projection, Editions Prodiex, Paris

[13] One way of explaining the famous Fantasmagoria effect, credited E.G. Robertson in the 1790's, is to see it as an attempt to increase the power of the Quinquet lamp (an improvement of an Argand lamp). The lantern was hidden behind the screen and rolled towards it on rails; the image seemed to approach the audience, simultaneously getting brighter. Of course this is not a full explanation; there were other contributing factors.

[14] Proust M (1954) A la Recherche du Temps Perdu I: Du Cöté de Chez Swann, Gallimard, Paris, p 12

[15] Something similar could be said about the early home movie systems like the Pathé Baby, introduced in the 1920's. The Pathé Baby projector used 9,5 mm film and a fairly weak projection lamp which did not allow a large screen projection. The framed Pathé Baby projection screen for the living room was hardly bigger than a regular television screen.

was epitomized exclusively by lantern projections would be incorrect. Naturally, there were images which did not need a special viewing apparatus: paintings, prints, photographs. An interesting tradition was the "Great Picture" genre – the business of touring with a large "sensational" oil painting, often in a sumptuous frame, which was shown against payment for the public.[16] There were also various popular apparata for consuming images both in the public and the private sphere. Most importantly, there was a remarkable proliferation of peep show devices. In such devices the image was hidden in a "box", to be looked at through some kind of a peep hole, often provided with a magnifying lens.[17] The image was not projected, but it had to be illuminated in some way; it was often transparent with the light coming from the back. All kinds of atmospheric effects (day to night etc.) were possible. The images shown in such devices spanned the whole range of visual techniques from painting and prints to stereoscopic photography, chronophotography and eventually film.

From a media-archaeological point of view it is significant that the first publicly exploited film-based moving picture device, Edison's Kinetoscope (1894) was a peep-show machine, not a projector. Although most film historians have considered the Kinetoscope merely a curiosity, an indecisive and failed attempt to find a viable form for the exploitation of "moving photographs", the prior existence of the tradition of the peep show box makes this understandable. Also Edison's idea of showing his Kinetoscopes collocated together in "Kinetoscope parlors" had predecessors. Edison had exploited his phonographs in a similar way; yet already from the early 19th century that there had been permanent peep show-based attractions, e.g. "Cosmorama saloons" (there was a big one in P.T. Barnum's American Museum in New York)[18]. Later public peep show attractions offered series of stereoscopic views arranged into travelogues for several viewers at a time; Aloys Polanecky's *Glas-Stereogramm-Salon* and August Fuhrmann's *Kaiser-Panorama* were cylindrical devices, surrounded by stereoscopic peep holes.[19] Visitors sat on chairs around the "panorama" and peeped at the three dimensional images which were changed automatically by a machinery. All these spectacles anticipated the Kinetoscope parlor.

Again, side by side with the public installations, the peep show devices found a place in the privacy of the home. In addition to various popular toys (such as the French *Polyorama Panoptique*), by far the most important domestic peep show

---

[16] See Cahn I (1996) The changing landscape of modernity: early film and America's 'great picture' tradition, Wide Angle 18 (3): 85–100.

[17] About the history of the peep show, see: Il mondo nuovo. Le meraviglie della visione dal '700 alla nascita del cinema, a cura di Carlo Alberto Zotti Minici, [Milano:]: Mazzotta, 1988; see also Füsslin G, Nekes W, Seitz W, Steckelings K-H W, Verwiebe B, Der Guckkasten. Einblick – Durchblick – Ausblick, Stuttgart: Füsslin Verlag, 1995.

[18] See an illustration of Barnum's "Cosmorama Department" in: Kunhardt PB Jr, Kunhardt PB III, Kunhardt, PW, P.T. Barnum. America's Greatest Showman, New York: Alfred A. Knopf, 1995, p.139.

[19] See Kiesinger E, Rauschgatt D (1995) Die Mobilisierung des Blicks. Eine Ausstellung zur Vor- und Frühgeschichte des Kinos, PVS Verleger, ipv., Wien, pp 51–58

device was the stereoscope, which first became a fashion in the 1850's and remained popular well into the 20th century. By its design the stereoscope excluded the viewer's immediate surroundings and focused his attention exclusively on the three dimensional view. In the Victorian society where the relationship between the outside world and the domestic sphere became increasingly polarized the stereoscope obtained an important mediating function.[20] It brought the outside world to the intimacy of the domestic parlour by means of stereo cards. The market for stereoscopic viewers and photographic stereo cards was huge. Geographic views, taken by travelling stereographers, were the most popular genre. Sometimes the stereo cards were arranged into systematic "world tours" with maps and guidebooks. Great numbers of them also concentrated on actual events; the effects of the earthquake in San Francisco in 1906 were a particularly popular topic. In these functions the stereoscope certainly preceded the future role of the television.

As a technical achievement and even as an object with formal features of its own the "small screen" does not stem directly from either of these traditions. Technically it was related to the development of the cathode ray tube, a device which was first successfully demonstrated around the turn of the 19th and the 20th century and became really practicable in the 1920's.[21] However, we can point out several intersections at which the prevailing ideas about the screen and the emerging notion of the small screen met and even merged with each other. A case in point, the television as an image box was obviously related to the earlier history of peep show boxes; however, to see it merely as a new kind of peep show device would be an oversimplification. Both on a discursive and on a material level the borders between different screen practices were far from clearcut during the late 19th century, the incubation time of the small screen.

Caught on the Act

Vilém Flusser has reflected on the relationship between the cinema and the television: "The movie theater is a late development of wall painting. This is its essence. The TV was projected to be a new type of window. It was meant to provide men with maps of the world to be used in subsequent commitments. That is what the word 'television' means: a better vision that [should read: than] is provided by conventional windows."[22] For Flusser, the big screen belongs to the realm of representation, whereas the small screen, epitomized by the television, is a melange between presentation and representation. Cinema combined the stabile monumentality of the wall painting with the linear modes of story-telling familiarized by the printing press. The small screen is not essentially a medium for telling stories

[20] See my "Armchair traveller on the Ford of Jordan". See also Pellerin D (1995) La photographie stéréoscopique sous le second Empire, Bibliothèque nationale de France, Paris

[21] See Jensen AG (1983) The evolution of modern television. In: Fielding R (ed) A technological history of motion pictures and television, University of California Press, Berlkeley and Los Angels, pp 235–249

[22] Flusser V (1997) Two approaches to the phenomenon, Television. In: Davis D, Simmons A (eds) The new television: a public/private art, MIT, Cambridge, MA , p 239

by means of pictures. It is an opening to observe events unfolding out in the world, a framed surface to see things that are beyond it: "One can see more of the world through it. Not only things that are too distant from conventional windows, but also things that are too small, or too ephemeral, or whose motion is too slow for conventional windows".[23]

As Flusser's analysis implies, when looking for pre-forms of the television screen, we should not separate form from function, the painted surface from the idea of mediation between two separate realms. However, from a media archaeological point of view the claim that a wall painting could never fulfill the presentational function assigned to television might not be correct. The images of an era are also defined by their relationships to other types of images. In the era of television oil paintings obviously do not have a presentational function; they are taken as representations. As various mythological stories testify this may have been the case in other cultures and in terms of other imaginaries. Narratives about people entering paintings and using them as gateways between different realms can be found from different parts of the world. Correspondingly, we may claim that images that are (with our norms) representations, mediated by pictorial techniques and the skills of the maker may have been conceived as providing much more direct "gateways" to distant realities.

This is one way of explaining the vogue for all kinds of peep show devices from the 17th century on. The peep shows were often exhibited by traveling showmen; they offered curious views for paying spectators. As the collector François Binetruy has stated, the great majority of the views shown in such devices were geographical views – cities, ports, palaces, courts.[24] For ordinary people whose possibilities to travel were very restricted, such views offered an opportunity for "virtual travel" (to adopt a modern concept). This aspect of the peep show culture was perhaps best expressed in the Italian name of the device, *Il Mondo Nuovo* ("The new world"). Particularly towards the end of the 18th century actualities (views about famous battles, catastrophes etc.) became more common. The peep show was a box which brought the world to its spectators; to nourish their growing thirst for images, there was a bulging industry for printed views, the most famous of which were the "Remondini" (from Bassono, Italy).[25]

The "reality effect" of the peep show was enhanced by various visual tricks. There were boxes which showed very deep perspective images, consisting of several picture planes (resembling the stage set of the baroque theatre). There were also ways of manipulating the amount of light in the box. By opening or closing a door behind the transparent image atmospheric effects could be produced. The backlight strangely anticipates the functioning of the cathode ray gun on electronic monitors. However impressive their effect may have been, the peep shows

---

[23] Flusser, Two approaches to the phenomenon, Television, pp 239–240

[24] In an interview in my television series, The Archaeology of the Moving Image, YLE (The Finnish Television), 1996, part 3.

[25] Carlo Alberto Zotti Minici, "Per una ricostruzione visiva delle vedute ottiche Remondini", Il mondo nuovo. Le meraviglie della visione dal '700 alla nascita del cinema, a cura di Carlo Alberto Zotti Minici, [Milano:]: Mazzotta, 1988, pp 205–239

were always (technically speaking) strictly "off-line" media. They showed represented scenes, which tried by illusionistic means to give the impression of immediacy, detail and movement.

There was another device which actually offered a visual "on-line" relationship to reality. The cultural role of this device, the camera obscura, has been largely misunderstood. In retrospect the camera obscura has been above all seen as the precursor of the photographic camera. In a sense the invention of photography meant discovering a way to make the images produced by a camera obscura permanent by optical-technical means. The fact that the camera obscura had been frequently used by artists as an aid for making sketches has supported this notion as well.[26] It was a device for capturing and freezing reality; that was all. The camera obscura, however, was seen differently by contemporaries. As Jonathan Crary has observed, it was assigned various different cultural interpretations.[27] We have several descriptions which show that people were often fascinated by its possibility of representing *movement*. The French jesuit Jean Leurechon wrote in his *Récréation mathématique* (1621): "Surtout, il y a du plaisir à voir le mouvement des oiseuaux, des hommes ou d'autres animaux, et le tremblement des plantes agitées au vent ..."[28]

As an apparatus camera obscura captured the view in front of its pinhole and transmitted it to the inside of the box. By means of mirrors this live image was often projected on a screen that was part of the apparatus. In large house-size camera obscuras the image from the outside was experienced on a "table", where it could be touched by one's hands. In small portable models the screen was often on the upper surface of the box. A transformation process took place (the perspective was flattened, image and sound were separated, only a framed sector of the outside reality could be seen at the time), yet for many contemporaries the camera obscura's fascination clearly resided in its "liveness", not in the freezing of the moment. By framing and displacing reality "caught on the act" this apparatus focused the viewer's attention. It anticipated the fascination with the future television screen, the enhanced window that Vilem Flusser described above.

The Tele-Mirror

The tradition of the magic mirror is another interesting media archaeological reference point. As Jurgis Baltrušaitis has shown, we can find numerous literary and visual references to magic mirrors from the 16th century on.[29] A regular mirror

[26] Kemp M (1990) The science of art. Optical themes in western art from Brunelleschi to Seurat, Yale University Press, New Haven and London, pp 196–197. Canaletto and probably Vermeer belonged to the users of the camera obscura; it must have contributed to the "magic realism" of their works.

[27] Crary J (1991) Techniques of the observer. On vision and modernity in the nineteenth century, MIT Press (an October Book), Cambridge, MA, p 29

[28] Cit. Mannoni, Le grand art de la lumière et de l'ombre, p. 23. Leurechon's description should be compared with the impressions triggered by the Lumière brothers' first film projections. The discursive similarities are thought-provoking.

[29] Baltrušaitis J (1978) Le miroir. Révélations, science-fiction et fallacies. Essai sur une légende scientifique, Elmayan/Le Seuil, Paris, pp 181–212

reflects a virtual space which is, however, a mere reflection of the physical space in front of it, even if it may be distorted. A magic mirror, on the other hand, reflects a reality which may take place simultaneously elsewhere or will take place in the future. Expressed anachronistically, it is used for surveillance – to monitor the doings of one's wife or loved one. More often than not, the devil intervenes and presents a view one really would not like to see.[30]

In Shakespeare's *Macbeth* there is a scene (Act IV, Scene I) in which the three witches show Macbeth a vision. The spirits of the kings of Scotland walk past him; seven have already gone, then an eighth comes holding a mirror (an early example of "portable technology"):

> *Macbeth:*
> "And yet the eighth appears, who bears a glass
> Which shows me many more; and some I see
> That two-fold balls and treble sceptres carry.
> Horrible sight! Now, I see, 't true;
> For the blood-bolter'd Banquo smiles upon me,
> And points at them for his."

Baltrušaitis has reproduced an engraving from the 17th century, showing a sorcerer conjuring up a vision of the future kings of France for Catherine de Medicis.[31] The kings are seen passing in a framed square mirror placed above the fireplace. Although the spectator, Catherine de Medicis, is standing, instead of sitting in front of the "screen", the placement of this "téléviseur catoptrique" (Baltrušaitis) seems to anticipate elements of future television culture. The magic mirror both looks and functions much like a television. The relationship between the "image window" and the fireplace is also noteworthy – it became an important discursive topic in the TV era.[32] Often the television was interpreted as having taken the place of the fireplace as the center of the domestic living room. It had became the "electronic hearth".[33]

How far should one go in drawing such parallels? Should we treat the conjuror as the precursor of the 20th century television director? Are the aliens appearing unexpectedly on TV monitors in science fiction movies somehow related to the

---

[30] Early cinema triggered a different but related topos. Both films (e.g. Ladislav Starewich's "The Cameraman's Revenge", Russia 1914) and cartoons often featured a situation in which a husband or a wife goes to the cinema and see his/her partner on the screen in amorous relations with someone else, captured by the relentless eye of the camera. Sometimes, as in Starewich's film, the cameraman played the role of the devil. See Bottomore S (1996) I want to see this Annie Mattygraph. A cartoon history of the coming of the movies, Le Giornate del Cinema Muto, Pordenone, p 59

[31] Baltrušaitis, Le miroir, p.187.

[32] Tichi C (1991) Electronic hearth. Creating an american television culture. Oxford University Press, New York, pp 42–61. During the early years of television, people sometimes places the TV set inside the hearth (p 45).

[33] Perhaps we should recall here the new age videotapes from the 1980's, which aimed at turning the TV set into a virtual fireplace by showing merely burning wood. And we should not forget the central role of fire in all kinds of magic rituals...

procession of the future kings of France?[34] It is difficult to point out any meaningful connection between 17th century "magic media" and the 20th century television, except perhaps in the common desire to see at a distance by means of "catoptric science" or "natural magic". And in fact the idea of conjuring up the future (media as a time machine) seems to have been more thrilling than extending one's vision in real-time by means of magic mirrors. The problem may be somewhat similar as looking for parallels between advances in telecommunications and the interest in telepathy (the word media is, of course, related to the the word "medium" in the parapsychological sense), or seeing the great popularity of stomach-speaking dummies in the 19th century as an anticipation of the disembodiment of the voice soon to be familiarized by the telephone. Such parallels are exciting but they are difficult to prove.

However, there have been other magic qualities assigned to mirrors. They have also served as gateways to alternative realities. The most famous instance of this topos is Lewis Carroll's *Through the Looking Glass* (1871), in which Alice enters a fantasy world by traversing a mirror:

> *Alice:*
> "Let's pretend the glass has got all soft like gauze, so that we can get through. Why, it is turning into a sort of mist now, I declare. It'll be easy enough to get through" – She was up on the chimney piece while she said this, though she hardly knew how she had got there. And certainly the glass was beginning to melt away, just like a bright silvery mist. In another moment Alice was through the glass and had jumped lightly down into the Looking-glass room.[35]

Here again the mirror is associated with the fireplace; the "bright silvery mist" connects the scene easily with earlier traditions of magic mirrors.[36] On the other hand the idea of the mirror as a gateway between different but related realms stems from the traditions of illusionistic painting as well, somewhat contradicting the notion of the painted surface as merely a surface – its interpretation obviously changes in time, in relation to other "image media". Be it how it may, this topos has manifested itself numerous times in the 20th century media culture from Jean Cocteau's poetic films *Le Sang d'un Poête* (1930) and *Orphée* (1950) to countless television advertisements, with the TV screen replacing the magic mirror. According to an early advertisement for TV broadcasting, the television set is a "looking glass" and the viewer a "modern Alice".[37]

---

[34] There seems to be a clear difference. The aliens usually address the perplexed TV viewer directly, whereas the figures appearing in a magic mirrors are unaware of the "presence" of the spectator. The spectator is a voyeur in the purest sense, just like the spectator of a peep show machine.

[35] Carroll L (1974) Alice's Adventures in Wonderland and Through the Looking Glass, Puffin Books, Harmondworth, Middlesex, p 195

[36] The apparitions in the magic mirror were often surrounded by smoke or a kind of mist. Similarly, projections of demons or spirits on smoke screens were one of the favourite techniques used by conjurors and magic lantern showmen in the 18th and the 19th century. See Levie F (1990) Etienne-Gaspard Robertson. La vie d'un fantasmagore, Le Préambule, Longueil, Quebec, pp 86–88

[37] Tichi, Electronic hearth, p.13.

When comparing the television set with a window, Vilém Flusser also considers the door: "The door is a hole in the wall which permits a rhythmic human motion: a diastolic phase in which man leaves himself to commit himself to the world, and a systolic one in which he comes to himself again without totally losing the world. The window, is however, a hole in the wall which provides man with a vision of the world which may serve as a map when he leaves the door to commit himself to the world. Thus the purpose of the window is linked with the purpose of the door, and that link has a dialectical aspect." [38]

Although this idea may hold for traditional television culture, in the case of more recent electronic screens the "window" and the "door" seem to be melting into one multifunctional opening. The display of a personal computer linked to the Internet is both a "window", a way of displaying information from all over the world, and a door for "entering" the virtual realms of CyberSpace. Here, however, the world that is observed and entered simultaneously has less and less to do with the world behind a traditional door and a window. It is a new surrogate reality which can only be accessed through this new opening. Alice's experience has frequently been invoked by advertisers in this context as well, perhaps better justified than in the context of television.

Telectroscopic Visions

The idea of *tele-vision*, "seeing at a distance", excited the 19th century technological imagination. It was undoubtedly influenced by advances in telecommunications throughout the century. A remarkable series of innovations took place: from the optical telegraph to the electric telegraph and to the telephone and the wireless. These, together with new means of transportation (the train, the steamship, and in the end of the century, the automobile) led to the notion that profound changes were taking place in the space-time continuum.[39] The globe was felt to be shrinking thanks to man's achievements. The romantic longing for elsewhere, colonialism and imperial expeditions as well as the positivistic science all gave their contributions to the increasing curiosity towards worlds "beyond the horizon". This curiosity was met by popular scientific literature, newspapers, lantern lectures, photography, stereoscopy and spectacles like the panoramas, which presented huge wraparound views of exotic geographic locations or famous events from the recent past (above all, battles) to an audience hungry for new visual impulses.[40]

Against this background it is not very surprising to encounter a device like the telectroscope. According to Carolyn Marvin, the telectroscope was "a popular but entirely imaginary invention of the late nineteenth century 'by which actual scenes are made visible to people hundreds of miles away from the spot'." [41] News about

---

[38] Flusser, Two approaches to the phenomenon, Television, pp 238–239

[39] Kern S (1983) The culture of time and space 1880–1918, Harvard University Press, Cambridge, MA

[40] About the panorama considered as a 19th century media, see Oettermann S (1980) Das Panorama. Die Geschichte eines Massenmediums, Syndikat, Frankfurt am Main

[41] Marvin C (1988) When old technologies were new: thinking about electric communication in the late nineteenth century, Oxford University Press, New York and Oxford, p 197

such "discursive inventions", which existed in the popular consciousness but not as real artifacts were circulated widely, particularly on the eve of industrial and world expositions. Even Thomas A. Edison was claimed to have invented such a device which would "increase the range of vision by hundreds of miles, so that, for instance, a man in New York could see the features of his friend in Boston with as much ease as he could see the performance on the stage".[42] Respected popular scientific journals like *Electrical Review* and the French *La Nature* published numerous descriptions of the telectroscope.

Telectroscope was a dream machine, an extrapolation of the possibilities of existing media.The idea was obviously inspired by certain features of the electric telegraph, above all the synchronization of the signal between the transmitting and the receiving end. Another impetus must have come from the knowledge that the transmission of still images over telegraph wires had been solved surprisingly early, by Bain in 1842 and Bakewell in 1847.[43] The first working system for transmitting images and texts (the predecessor of both the telephoto and the telefax) was the Pantelegraph invented by Abbé Caselli in the 1850's and used commercially in the next decade, mostly for verifying signatures and distributing images of sought-after criminals.[44] Although these systems certainly belong to the prehistory of television, they were meant, like the ordinary telegraph, for communicating messages from one place to another. They did not directly tackle the idea of "seeing at a distance" or "seeing by electricity", which occupied numerous researchers during the second half of the 19th century.[45] Fantasies about the telectroscope had some basis in actual research, although concrete results were not attained until the turn of the century.

Towards the end of the 19th century the telectroscope (sometimes called "teleoscope") entered the public imagination in caricatures and science fiction stories. Edward Bellamy's prophesy about its use in his novel *Equality* (1897) was typical: "You stay at home and send your eyes and ears abroad for you. Wherever the electric connection is carried ... it is possible in slippers and dressing gown for the dweller to take his choice of the public entertainments given that day in every city of the earth."[46] This vision gets surprising close to the cultural form that eventually came to dominate radio and later television: broadcasting. The world is described as a source for potentially endless channels of entertainment. The role of the user is restricted to tuning and enjoying the global program feed from one's living room. Such a vision was depicted also visually, for example by Albert Robida in his book *Le Vingtième siècle* (1883). Robida shows an elderly man leaning back in his armchair, smoking a cigar while watching and listening (!) to a performance

---

[42] Electrical Review, May 25, 1889, p.6. Cit. Marvin, When Old Technologies Were New, p. 197.

[43] Hogan JVL (1983) The early days of television. In: Fielding R (eds) A technological history of motion pictures and television, University of California Press, Berlkeley and Los Angels, p 230

[44] Couchot E (1988) Images. De l'optique au numérique, Hermes, Paris, pp 75–79

[45] For an overview, see Lankes LR (1983) Historical sketch of television's progress. In: Fielding R (eds) A technological history of motion pictures and television, University of California Press, Berlkeley and Los Angels, pp 227–229

[46] Cit. Marvin, When old technologies were new, p 200

of cancan transmitted by a "telephonoscope".[47] As the name of this fictitious apparatus reveals, combining the transmission of image and sound, Robida had been probably influenced by experiments in transmitting regular ("radio") programs by means of the telephone.[48]

In the late 19th and early 20th century visions about the telectroscope the idea of "broadcasting before broadcasting" was not the only option. Often the device was conceived as a personal tele-vision device, handled actively by the spectator almost like some kind of an electric telescope. In Harry Grant Dart's illustration for *Life* (1911) we see an elderly man in his parlor, surrounded by all kinds of fantastic gadgets.[49] He is currently using an "Observiscope", another variant of the telectroscope. By pressing a button the person in question can select a view from a menu; currently we see him observing his son Willie embracing a girl. In this pessimistic vision, sarcastically titled "We'll all be happy then", tele-vision serves the purpose of voyerism and surveillance: the couple on the display are seen from behind their backs. In other, more optimistic visual fantasies the television device serves as an audiovisual communication device between people, an early predecessor of the picturephone. In a well-known illustration from *Punch* (1879) we see an elderly couple sitting in front of a screen, discussing by means of a telephone with a young lady visible on the screen.[50]

From the Keyhole to the Screen

The preceding discussion shows that although the telectroscope certainly anticipated the television, its proposed use was not restricted merely to the function which later became known as broadcasting. The idea of "seeing by electricity" was an open concept, which could be applied to various purposes. What was the apparatus and particularly its display like? Most visual discourses I have found depict only the user's end. Just like its proposed functions, also the configuration of the apparatus varies. The most fantastic proposal seems Albert Robida's idea of public television (1883). The events of the world are transmitted live on huge oval or

[47] See Canto C, Faliu O (1993) The history of the future. Images of the 21st century, Flammarion, Paris, p 32

[48] In his earlier novel Looking Backward (1887) Edward Bellamy had described a home music room. As Michael Brian Schiffer explains, "After consulting a program that listed the day's offerings, the listener adjusted 'one or two screws', which filled the room with music 'perfectly rendered.' The program came to every home via telephone from central music halls where the best musicians performed twenty-four hours a day. On Sunday mornings, there was even a choice of sermons. " (Schiffer MB (1991) The portable radio in American life, The University of Arizona Press, Tucson London, p 12. About actual such systems in the late 19th century, see Marvin C (1991) Early uses of the telephone. In: Crowley D, Heyer P (eds) Communication in history, Longman, New York London, pp 145–152

[49] Reproduced as a frontispiece to my article "From Kaleidoscomaniac to Cybernerd. Notes Toward an Archeology of Media", Electronic Culture, p 296

[50] See Toulet E (1988) Cinématographe, invention du siècle, Découvertes Gallimard and Réunion des musées nationaux, Paris, p 47. The wide screen which is placed above the fireplace doubles exactly the position of the magique mirror referred to in note 31.

round screens displayed on the outside .[51] The enormous screens, resembling giant crystal balls, are watched by audiences from the balconies of a nearby building. Although public television viewing has never made a real breakthrough, Robida's intuition seems to anticipate the giant videoscreens at stadiums or in urban spaces like Tokyo. It also shows that the dynamics of private/public, the giant/the tiny observed in the context of the magic lantern in the 19th century also manifested itself in the fantasies about "tele-vision".

Recalling the dominance of the two different visual apparata, the magic lantern and the peep show box (described earlier), in the 19th century, it is perhaps not very surprising to find out that both of them influenced the design of the telectroscope. In illustrations the telectroscope is the most often imagined as a kind of tele-magic lantern; the image it transmits is projected on a wall or a screen. Almost without an exception the image is round, just like many of the projected lantern images. The device itself may be placed conveniently on a table in front of the user, as in the cartoon "We'll all be happy then" described above.[52] Interestingly, a transformed magic lantern was really sometimes used as the television receiver in early television experiments, like those by the French inventor Eduard Belin.[53] Leaping across time this tradition could also be compared with the current interest in "home theatres", where the television set has been replaced by a video projector, a late inheritor of the magic lantern.

Another variant was the telectroscope as a table top peep show machine, clearly influenced by the design of the common parlor stereoscope. In an illustration from 1890 we see a man peeping into such a device while an Asian servant waits behind his back, holding a tray; a helmet and a whip placed on a chair identify the man as a colonial officer, probably peering into his distant homeland (no "microphone" is visible).[54] That the tele-vision device should probe the form of a familiar domestic peep show device makes as much sense as its relationship to the magic lantern, perhaps even more. Although the stereoscope was a purely optical device, its role in the domestic sphere anticipated the future role of television, as has already been explained. Even without being "wired", the stereoscope was used to gaze at a distance; the enhanced reality effect of the three-dimensional visual space compensated for the lack of real-time communication.

A missing link, in more than one sense, between the stereoscope and the television set was the Kinora, an early table top device for viewing moving photo-

---

[51] See the reproductions in Zdenek Krecan (1979) Encyclopédie illustrée de l'image et du son, Gründ, Paris, p 330 and 359

[52] Another example is provided by a publicity card for Lombart chocolate from the series *En l'an 2012* (Imprimerie Norgeu, France 1912). A woman is sitting by her desk, talking on the phone to a man whose image is projected on the wall by a "tele-lantern" placed on the desk. The transformer has been placed on the floor by its side. The man on the screen seems to be a colonial officer in a Far Asian country. Reproduced in Canto and Faliu, The history of the future, p 46

[53] See a photograph of Belin's early equipment in Prima del Cinema. Le Lanterne Magiche. La collezione Minici Zotti, a cura di C. Alberto Zotti Minici, Venezia: Cataloghi Marsilio, 1988, p. 95.

[54] Reproduced, with no source mentioned, in: Kloss A (1987) Von der Electricität zur Elektrizität, Birkhäuser Verlag, Basel, p 245

graphs in the home.[55] Instead of celluloid film, it used flip book-like reels of paper-strips, with one film frame copied on each; these reels were handcranked by the user. The Kinora, patented by the Lumière brothers (the main proponents for film projection!) in 1896 was another peep show device, which adopted its basic form directly from the stereoscope; the 3D illusion was replaced by the thrill of the moving image. Interestingly, there were more expensive De Luxe models for three, even five people at a time. Instead of peep holes, these models had bigger square-shaped magnifying lenses mounted side by side. These Kinoras meant that the viewers no longer had to immerse their eyes completely into the "keyhole", nor did they have to watch the images alone. With Kinora, which went out of production well before the 1920's, the idea of the small screen was clearly in the making, at least unknowingly.

If we look today at the early experimental television sets from the 1920's and the 1930's, for example those designed by the Scottish Baird, the German von Ardenne or the American Farnsworth, we are struck by the small size of the screen. Indeed, it is hardly larger than lens of the De Luxe Kinora and certainly smaller than the magnifying lenses of most traditional peep show machines. In addition the screen is often round, again like in peep show machines. Photos of early television spectators often show them gathered around the television set, leaning towards the screen. Their eyes are no longer immersed into a peep hole, but they are certainly much closer to the image than in a magic lantern projection. This proxemity may be explained by the aims of the marketing people, as well as by practical realities. The image was not only small, its resolution was poor.

Why was the image so small? The most concrete explanation is technical. There were two competing television systems, mechanical and electronic. In the former (stemming from the patent applied by the German Paul Nipkow in 1884) the image was both deconstructed and then reconstructed (after transmission) by means of a slotted scanning disc spinning in front of a light sensitive cell.[56] For practical reasons (to limit the size of the apparatus) the image was deemed to remain small, really a kind of peep show image.[57] In the electronic system the image size was dependent on the development of the cathode ray tube which was invented around the turn of the last century. The size of the cathode ray tube and consequently of the screen increased steadily but slowly as the design and

---

[55] See Anthony B (1996) The Kinora. Motion pictures for the home 1896–1914. The Projection Box, London. Different Kinora models are shown on pages 36–39.

[56] The French magazine *La Nature* published in 1898 a description of "Monsieur Dussaud's Teleoscope", which was based on the design of the Nipkow disc. As had happened before, the magazine promised that "Monsieur Dussaud is at present engaged in improving his system, and it is hoped that he will succeed in completing his experiments in time for the Exhibition to be held in Paris in 1900." Before that Mr. Dussaud seems to have disappeared from history. De Vries L, Victorian Inventions, London: John Murray, 1971, pp 141–142

[57] An intriquing thing is the close formal resembrance between Nipkow's disc and some "pre-cinematic" devices, such as Georges Demenÿ's Phonoscope and Ottomar Anschütz's Electrotachyscope, conceived at the same time. Both Demenÿ and Anschütz used a spinning disc. The images were viewed through a peep hole, although even a projecting version was available.

manufacturing methods improved. Screens approximating the sizes of contemporary television screens became possible only in the late 1930's.

Digging Deeper

The nature of the cathode ray tube effected the design of early television sets in other ways as well. The bigger and better tubes of the 1930's often necessitated a very long cathode ray gun. Consequently also the tubes became very long (the opposite of today's flat liquid crystal displays). This was inconvenient from the point of view of the television design. An interesting, albeit today a weird looking solution was found. The tube was placed inside the television set vertically, the screen facing upwards. It was hidden by a cover, which had to be opened before the viewing session could start. There was a mirror on the inside of the cover, which was left in a slight angle. The screen itself was invisible for the spectator and the television image was watched as a reflection in the mirror.

Although such an explanation makes perfectly sense, it is not sufficient for the media archaeologist, who endeavours to dig deeper underneath such matter-of-fact explanations. It is highly intriquing that exactly the same arrangement of the image and the mirror had actually been used in countless peep show machines already hundreds of years earlier. In such devices the image was placed horizontally at the bottom; a mirror placed at 45 degrees above it reflected the image to the peep hole, thus emphasizing the perspective illusion. One might also want to recall the tradition of the magic mirrors. With the introduction of the television sets of the 1930's the mirror was once again adopted as a screen for "viewing at a distance". Are such parallels merely coincidences? Or are there some hidden connections to uncover beneath the surface of technological development?

Without aiming at conclusiveness, this article has purported to show that the early development of the small screen was not determined merely by technical factors. These factors themselves are always embedded in larger cultural processes. The idea of "seeing at a distance" by means of technology had been a cultural presence well before it existed as an artifact. It had been exposed to other ideas which had been widely distributed and deeply rooted in various mentalities. In the 19th century there were countless conflicting and overlapping discourses related to the idea of the screen. It wasn't clear that the tele-vision machine would look like a piece of furniture with picture window on its side. Neither is it clear that it will always have to remain that way. Indeed, as Margaret Morse has observed, "the television receiver as box is gradually being displaced by the computer display terminal, integrated with the telephone, the typewriter and drawing tablet, the library, the game and bulletin board".[58] But does it mean that we are now finally entering the zone of the unique and the unprecedented? I doubt it.

---

[58] Morse M (1990) The end of the television receiver. In: Geller M (eds) From receiver to remote control: the TV set, The New Museum of Contemporary Art, New York, p 139

The Technoetic Dimension of Art

Roy Ascott

Classically, art has been complicit in reinforcing authorized reality. How tirelessly Renaissance artists toiled at the fabrication of their rule-based perspective. How endlessly the crude materialist message was reiterated over the dark Northern centuries of 'humanistic' representation and laundered narrative. Only Cézanne had the vision and enlightenment to break through that hardened shell of reality into the shifting, fragmented contours of a world in flux. It was Cézanne, at the turn of the century, who first understood, intuitively, poetically, that art like reality was a negotiation; the product, never finite, of interactions and transformations within the realm of consciousness . This realm, we have come to understand, includes both the observer and the observed, and constitutes a dance of meaning and perception, in which no aspect is stable, no relationship definitive, no process complete, no system closed or indefinitely looped. And it was not until forty or fifty years later that the full metaphysical implications of Cézanne's vision were realized in the warp and weft of Pollock's visual networks. Cézanne understood the painting as an organism that could evolve in the way that the landscape of Mont St. Victoire evolved, through the interaction of forces. Pollock brought this principle of interactivity to full term by placing his own behaviour as the very subject and centre of the constructive process.

Both artists and scientists today, led by sheer empiricism, increasingly proceed on the basis that reality is constructed. With some surprise we are beginning to realize that the constructing system is ourselves. The paradox is that the mind that makes our reality is itself part of the reality it makes. There is a structural coupling, as Maturana and Varela have shown: we are self-creating, our reality is a part of the autopoietic process.[1] To the classification system which we apply to our processes of intelligent awareness and action, we can add a new human faculty, which I call cyberception,[2] involving computer augmented and telematically enhanced cognition and perception. Cyberception is enabling us to understand the role of our consciousness in emergent reality while, at the same time, engendering a new sense of Self and a new definition of human identity. We live increasingly between the virtual and what we have become accustomed to call the real. It is in part due to our ability within this hybrid space not just to conceive of alternative worlds but to actually engage in their creation that we see nature itself as susceptible to reconstruction and re-articulation, particularly that part with which we co-evolve.

All our instincts now are towards the construction of new realities rather than the representation, reiteration, interpretation or expression of worldviews which have been laid upon us. It has been a long process. Both religious duplicity and materialistic determinism have held tenaciously onto our vision, clouding and blinkering our understanding. It has taken most of this century, searching, fumbling in many cases, to find our way into the deeper mysteries of the Real, to understanding the artist's role in making the invisible visible, of opening up processes of emergence and complexity. At century's end, we might say at the millennium's end, we are understanding that art can more fruitfully be concerned with apparition than appearance, dealing more with processes of coming-into-being than with reaffirming the given. Philosophers see this as a Radical Constructivism, we artists see it as a radical connectivism. In little more than a decade, connectivity has become an irreducible, fundamental dimension of experience along with space and time. Our measurement and coordination of events must now be constructed along three axes, with connectivity added to temporal duration and spatial extension.

In looking at the importance of science for art, I turn to my own development as an artist, which, following my university research into *Cézanne and the expression of change*, started in 1959 with constructions such as "change paintings" and "hinged reliefs" which were fabricated out of plastic, glass and wood and consisted of sliding, folding, interlocking panels, often transparent bearing images, which were so constructed as to invite the viewer to manipulate the parts and thereby "create" the work of art, and in effect create its meaning. My first London one man show in 1964 was entitled *Diagram Boxes and Analogue Structures*. At that time I developed a theory of participation and interaction based on my study of cybernetics and in 1964 I published *Behaviorist Art and the Cybernetic Vision.*[3] My research was deeply invested in the works of Ross Ashby, F. H. George, Stafford Beer, and of course Norbert Wiener. I started a two year "Groundcourse" at a London art school,[4] radically modeled on cybernetic principles, behavioral psychology, and the process philosophy of Whitehead. Throughout the sixties I consorted with scientists in Britain, Jacob Bronowski bought my work. Frank Malina recruited me to write for his new journal of art, science and technology, LEONARDO, whose editorial board I have served on ever since. At this time I published a manifesto which spelt out in many ways the prospective for art as it developed over subsequent decades (see Appendix).

In California in the 1970s, introduced to the computer conferencing system of Jacques Vallée, *Infomedia*, I saw at once its potential as a medium for art and in 1979 abandoned painting entirely in order to devote myself wholly and exclusively to exploring telematics as a medium for art, or as it is now called Art in the Net. Since that time, the shift in the technology of telematics alone has been exponential, as my own experience shows. When I initiated my first international telematics project *Terminal Art* in 1980, the infobahn was little more than a dirt track. I had a Texas Instrument portable with the old rubber acoustic couplers linking me by phone to Vallée's worldwide Infomedia network. I mailed out a dozen such terminals to artists in various parts of the States and Europe, including Douglas

Huebler and Eleanor Antin in California, and Douglas Davis in New York. There was only paper printout with no screen, and of course images were out of the question.

Three years later at Frank Popper's *Electra* exhibition at the Museum of Modern Art in Paris, I initiated *La Plissure du Texte, a planetary fairy tale*[5] with archetypal characters (groups of artists) at computer network nodes contributing to a nonlinear narrative from some fourteen different geographical sites and time zones around the world. This kind of textual "exquisite corpse" had the "Prince" online at a node in deepest Quebec, the "Fairy Godmother" performing nightly at the Art Gallery of New South Wales, and the "Wise Old Man" logging on from the beach at Honolulu. This time we had screens, data projectors, the I.P.Sharp worldwide network. But still only ascii text. It was a deeply layered text, in turn witty, profound, critical, poetic.

Two years on, as International Commissioner for the Venice Biennale of 1986, I organized remotely through the same I.P.Sharp network, a large and complex exhibition of interactive computer installations and worldwide telematic projects. Apart from a couple of face to face meetings with my colleagues in Venice, the exhibition, involving over a hundred artists, was organized in dataspace by means of the same old and much traveled portable T.I. terminals.

By 1989 a far greater range of computerized telecommunication possibilities could be harnessed. My project for Ars Electronica in Linz, *Aspects of Gaia: digital pathways across the whole earth,*[6] used electronic images, sound, and text sent by all possible means between all possible places, through computer networks, bulletin boards, slowscan TV, fax, even radio transmission. An electronically controlled trolley wafted the visitor through a darkened tunnel full of scrolling LED signs and music bites, carrying poetry from all over the planet.

We rafted into the Nineties on the PowerBook, plugging into the Net wherever we went. Compression made image exchange easier. Quicktime software put video into circulation. Refresh rates got faster, bandwidth got wider, thought never fast or wide enough. When in 1993 Alex Adriaansen asked me to do a project for V2, I chose to limit it to a solid twenty-four hour period, with wall to wall communications technology of all kinds, digital and analogue, squeezed in between Saturday and Sunday, to create the eight day of the week, the day of *Telenoia,* a celebration of connectivity and mind-at-large.[7]

Now Art in the Net is an established form, the telematic substrate providing the base of work which can only originate, exist, and develop in a network environment. My most recent project, conceived, researched, created and developed wholly online, *Identity in Cyberspace,*[8] was truly collaborative and cooperative, involving remotely distributed artists in Spain, Ireland and Wales. Significantly, Ars Electronica in 1996 dropped its computer graphics award category in favor of Web design. In November, ICC had the first online art exhibition of highly diversified works created entirely by, in and for the Net.

The case for telematics as an art form, with its own practice, theory and history, is now established. But this is simply the very beginning of what can become an enormously rich and fruitful artistic endeavor. Telematics is all about mind, it involves the technology of consciousness, what I describe as the technoetic principle (noetic is derived from Greek "nous", mind). Technoetics brings into our experience a fourth irreducible quality of existence, to compliment, as it were, space and time and connectivity. This is the dimension of consciousness.[9]

Without the connectivity, the "hyperlinks" in multimedia terminology, of widely ranging associative links and conceptual interaction, our experience of space and time is null and void. This is to speak of mind-at-large, mind unfolding, blossoming within the net. It is telematic mind, the collective intelligence of which is reaching that level of complexity which augurs the emergence of a kind of hypercortex within our evolutionary flow. New collaborative strategies, new interactive behaviors working through the medium of the Internet, enrich the connectivity, heighten the *interneticity*, of this hypercortex. It is the hypercortex which is now at the core of our reality engine. Heinz von Forster concluded his seminal essay *'On constructing a reality'* with the simple formula: reality = community. The development from there is clear. Our community is telematic community, society online. Our social behaviour is virtual behaviour. Art now occupies the domain of a radical connectivism, in which the transformation of consciousness is the primary condition of our telematic culture. The consequences of a hypercortex for the future of art are likely to be quite unprecedented. We can foresee the importance of the hyperconcept, that is the layered idea which exists always and only in a constellation of associative links of extreme subtlety.

The hypercortex will at once inform and be informed by the structural coupling of each individual within the network, and thereby affect the status of human identity leading in evolutionary terms to the emergence of the 'subtle being'; the Self as a much more negotiable, transactional, impermanent, transformative being. The Self as an ongoing creation, open to differentiation, distribution and dissemination through telepresence, in cyberspace, a kind of non linear identity in which fixed patterns of behaviour, conditioned responses, learned protocol become the exception rather than, as in the classical western model, the rule. Such subtlety of being will call for a 'subtle art'. This 'subtle art' will embody an 'interstitial practice', art located at the intersections of cognitive science, bio engineering, and metaphysics.

However, it is no longer enough to speak of the convergence or reciprocity of art and science in some general sense, that old modernist dream, but to specify which art and which science, and by what means they might fruitfully interact. What ultimately we are looking at is an art of artificial life and a science of dynamic systems, interacting within the complexities of a telematic culture. While artificial life, nanotechnology, molecular engineering promise the eventual re-materialization of art, after this post modernist era of digitally driven, screen based dematerialization of art, its real significance to art in the new millennium will ultimately be at the level of mind. That is to say that artificial life, radical a practice as it appears in these early days of its development, will only become of consequence culturally when it gives rise to artificial mind, prosthetic consciousness. This is to speak of course to something qualitatively much more than 'artificial intelligence'.

So, no matter how advanced or sophisticated artificial life technology becomes, no matter how it might impact formally on the constructive power or repertoire of art practice, and I have no doubt that it represents the most important road that art at the turn of the millennium can follow, it will fail absolutely to realize its promise as the agency of great creative force if it is not embedded in continuing research into consciousness. It is not the noetic principle which is paramount here, so much as the technoetic principle. To understand the full potential of telematic mind and artificial consciousness, to advance a truly technoetic aesthetic, to open up the conceptual and psychic riches of a telematic culture, it will be instructive to look beyond technology, beyond western metaphysics, right back in time, across cultures, deep into our own internal histories. I am proposing that we need to examine closely the dimensions of a quite other and apparently alien space of the mind, the consciousness and culture of the shaman.

In relating the shamanic tradition of the old world to the aspirations of our telematic culture, we may raise useful questions concerning the architecture of contemporary life. Where the ancients were nomadic, we are restlessly 'telemadic', our minds traversing the vast interspaces of the worldwide networks of technology and consciousness. The shaman inhabits the interspace between this world and other realities, while we, both philosophically and technologically, are learning to understand what it is to construct reality, to monitor our own self-creation in the interspace between the virtual and the real. The old wisdom which invoked the cosmic rule "as above so below" can equally be applied to the modern world in demonstrating the relationship between immaterial life in cyberspace and the materiality of artificial life. "As in cyberspace so in material space". The recent successful cloning from a sheep's single gene, permitting the same, identical clones to be physically present in many locations at the same time, is the equivalent in material terms of telepresence in cyberspace, where one individual can be present in many locations at one and the same time. We should perhaps define a generalized theory of *teleclonics* to cover ideas of distributed presence, both de-materialized and re-materialized!

Those of us addicted to mind-altering technologies may have much to learn from shamanic culture, just as some shaman-led societies may be surprised to learn of the emphasis on consciousness that dominates current scientific research. This I intend to discover with an expedition into the Amazonian rain forest to meet with Xingu Indians.[10] I shall exchange experiences in CyberSpace with shamanic experiences of psychic space. Here, respectfully, may be room for cultural convergence. Where entities unknown to contemporary western society populate the psychic landscape of shamanic journeys, entities undreamed of in the ancient world, are emerging as life-forms within the Net and will eventually materialize in the nanomoisture of wet biology. The key questions facing us right now are: how do we house the global mind, how do we interface with other realities, how shall we replace the dichotomy of artifice and nature with the seamless connectivity of a technoetic ecology? As we seek ways to live coherently and creatively in cyberspace with the disembodied, virtual presences of the artificial living and the re-embodied intelligences of agents and avatars, what could we learn from meetings with shamanic personalities and our deep immersion in their psychic space?

Nothing impels us to creativity so much as our desire to understand what constitutes life. And all study of living process entails a study of consciousness. Where the renaissance led the artist to treat the human body as the apotheosis of the living being and to celebrate its triumphant materialism in an almost obsessive concern with the outward and surface aspects of form and structure, the twentieth century by and large has seen the artist attempting to explore the hidden and silent dynamics of life at the microscopic if not molecular level. We do not doubt that an immense number of elementary phenomena still escape our observation, and by the inevitable finite nature of our observing systems, perhaps always will do so. There will be a *quid proprium* of life which will always be irreducible. In order to understand the intertwining warp and weft of mind that characterises the tissue of telematic art, I like to take a wholly material analogy from the "culture of tissues". Now, the biology of tissue may not seem at first sight to be exemplary to an understanding of the state of development of the digital arts, and yet the recitation of a remarkable experiment first made by Alexis Carrel over sixty years ago bears some consideration and reflection:

> "If two palpitating fragments, proceeding from the same heart (of a dead creature), are put next to each other , but without touching, one usually observes that their rhythm is not identical. One fragment beats eighty times a minute, for

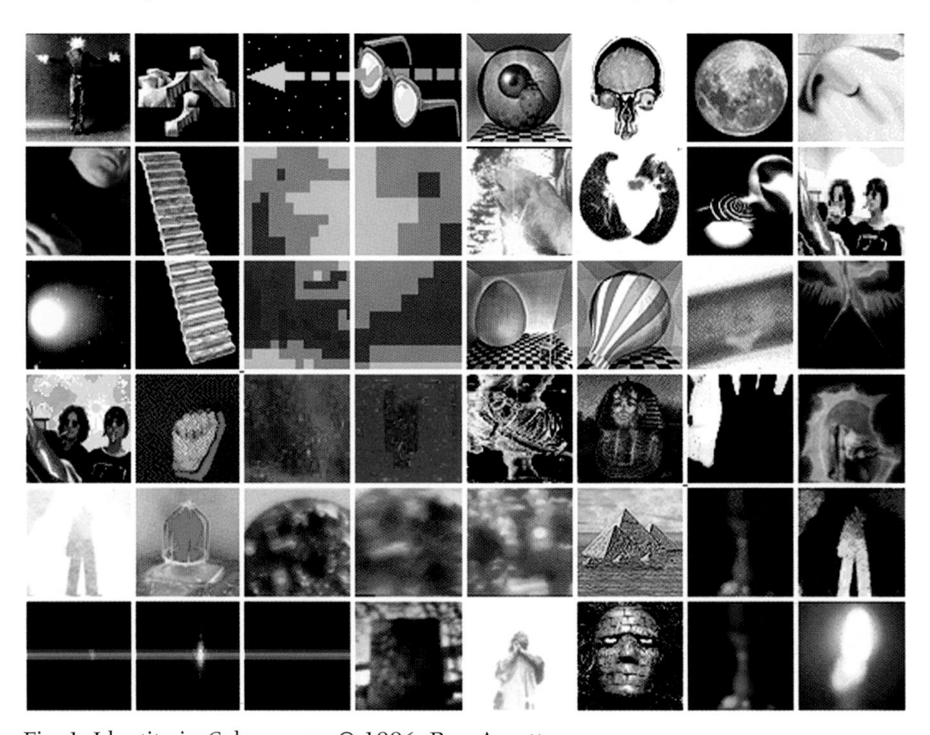

Fig. 1. Identity in Cyberspace. © 1996, Roy Ascott

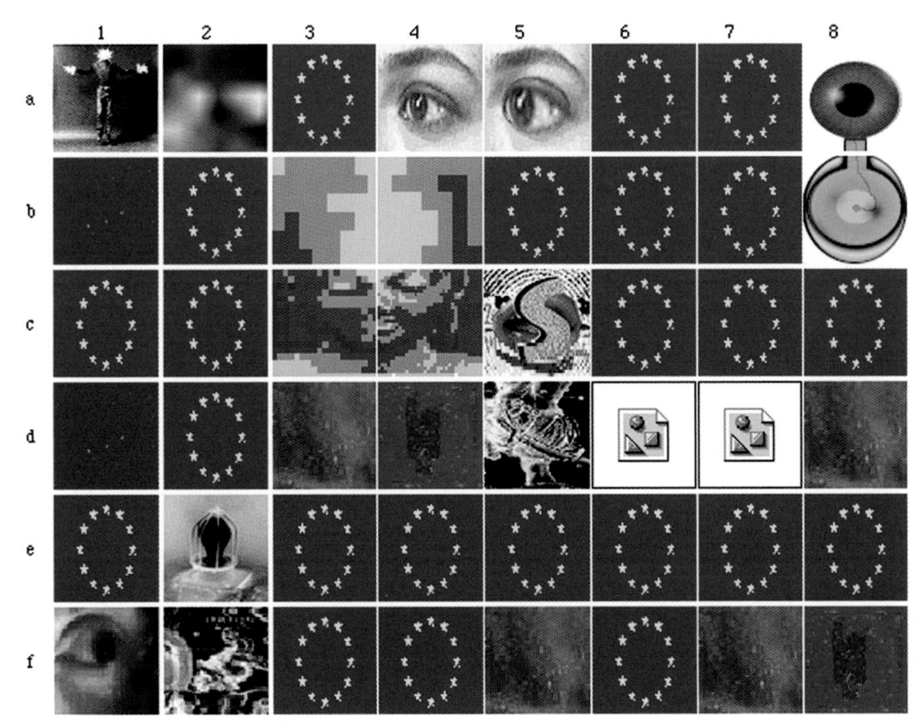

Fig. 2. Identity in Cyberspace. © 1996, Roy Ascott

instance, the other fifty. If it so happens, which is rare, that their pulsations are identical, they are nevertheless not synchronous. But these fragments proliferate and gradually surround themselves with a circle of new cells which penetrate into the medium in the shape of a thin, translucid, living layer. After a certain length of time these membranes issued from the two fragments come into contact. At that very moment, *the rhythm becomes identical.* The synchronism is reestablished: the two fragments beat as one. The bird has long been dead; the small pieces of muscle separated from its heart have neither blood circulation nor a nervous system connected to the main trunk, and yet they seem to recognise each other as soon as they come into contact."[11]

I think such identical rhythm, synchronicity and mutual recognition are to be sought as features of telematic life of technoetic culture. But the connection to living tissue is more forcibly evident and relevant in that special case of technoetic culture, Artificial Life. AL is significant in its materiality in respect particularly of its implications for the environment, for architecture, for the accouterments or cyburban living. AL means intelligence inhabiting artificially autopoeitic entities with which we might fruitfully cohabit. AL art models these relationships, previsions our mutual association, and imbues this future with a poetic dimension. Post biological, hyper technological living is always presented as epic, heroic or tragic. Instead, artists such as Christa Sommerer and Laurent Mignonneau, or

Thomas S. Ray for example, bring a lyricism, a flowing poetry to their work with this technology that raises not only its conceptual horizon but its human value.

So within the *telematic embrace*[12] we hope to reach this social synchronicity, an identical rhythm of the heart, a dynamic technoetic equilibrium. The meaning of hypermedia is to be found not in the collisions or collusions of content, not even the re-creation of meaning, seen as the semantic resuscitation of narrative after the banality of post modernist desecration. No, the meaning of hypermedia is ultimately spiritual, working at the level of a kind of associative consciousness, an interpenetration of minds or, better put, the seamless merging of mind fields.

But for all the richness of interactive, telematic projects at the poetic and sometimes visionary level of practice, the history of digital art by and large has been the history of craft. We have been living through a kind of Arts and Crafts movement not unlike that of the 19th century. Formal conceits, decorative flourishes, fractal flashiness have constituted the commonality of practice. In the evolving discourse of digital art the line between hype and necessity has been a thin one. Similarly in science, despite the best intentions, and great technological sophistication, we can only scratch at the surface of events. All our phenomena are 'envelope' phenomena, with an immense number of elementary phenomena escaping our observation. The problem is not the degree of resolution, the analytical acuity of our investigative instruments or experimental apparatus, instead it lies in the degree of granularity, as it were, of our own minds. The subtlety of thought required to penetrate further into the nature of reality, depends on a dynamic cognition of greater complexity than the singular human mind can sustain. It is to the hypercortex that we must look for the evolution of our intellectual and creative powers.

To aid this evolution, and implicitly or explicitly that is the role of the artist at this time, we neither wish to work in isolation nor can materially afford to do so. Much is to be researched, much to be discussed, tested and developed. For this reason we have seen research centres springing up in recent years to bring into cooperation technological and intellectual resources, from the domains of science and philosophy as well as art. ATR in Kyoto, ICC in Tokyo, Media Lab in Boston, ZKM in Karlsruhe, Ars Electronica Center, Linz, each in their different ways have extended the horizon for art while finding in their research new systems, strategies and concepts that have value in other disciplines and contexts, whether scientific, commercial or industrial. There is a brisk *quid pro quo* that puts the artist in a space far removed from the ivory tower and close to the hub of radical thought which is shaping the future and informing our destiny.

A relative newcomer to this nexus of research centres is CAiiA at the University of Wales, Newport. CAiiA, the Centre for Advanced Inquiry in the Interactive Arts, which I founded in 1994, has already signaled the initial diversity and range of its inquiry by the nature of the artists who are registered in its doctoral program[13] and the depth and variety of their projects and practices. CAiiA is a virtual community,[14] with each of us dispersed for long periods of time around the world, interacting through the Net, across our Web sites, pursuing research within the technoetic dimension. We can see virtual communities spreading around the world such that a Planetary Collegium[15] may properly be proposed as the future source and sum of advanced research in the technoetic domain.

Such is the complexity in technoetic art of the relationships between mind, behavior, environment and technology that research is now at the basis of many artists' practice. In this we use scientific method but are not bound by it; we employ new technology but are not constrained by it; we engage in theoretical discourse but are not directed by it. We have come a long way in the 20th century from the aesthetic of appearance[16] and expressive representation. But we should not forget our psychic roots. Art is an activity of mind before it is embedded in technology. Technology is an expression of the spirit before it is determined as utility. The technoetic dimension defines the trajectory that art will take as it propels us into the new millennium.

Appendix – Manifesto of 1968

Behaviourables and Futuribles

Roy Ascott 1968

When art is a form of behaviour, software predominates over hardware in the creative sphere. Process replaces product in importance, just as system supersedes structure.

Consider the art object in its total process: a behaviourable in its history, a futurible in its structure, a trigger in its effect.

Ritual creates a unity of mood. We need a grand rite of passage to take us from this fag end of the machine age into the fresh new world of the cybernetic era.

Just as our environment is becoming more and more automatic, so our habitually automatic behaviour becomes less taken for granted and more conscious and examined.

Now that we see that the world is all process, constant change, we are less surprised to discover that our art is all about process too. We recognise process at the human level as behaviour, and we are beginning to understand art now as being essentially behaviourist.

Object-hustlers! Reduce your anxiety! Process culture and behaviourist art need not mean the end of the Object, as long as it means the beginning of new values for art. Maybe the behaviourist art object will come to be read like the palm of your hand. Instead of figuration – prefiguration: the delineation of futuribles. Pictomancy – the palmistry of paintings – divination of possible futures by structural analysis. Art as apparition? Parapsychology as a Courtauld credit?

Cézanne's structuralism reflected a world flooded with physical data. Our world is flooded with behavioural data. How does that grab you?

Social inquisitiveness is a factor we would like to reinforce.

All in all we are still bound up with the search for myths. But the context will be biological and behavioural – zooming through the micro/macro levels. Get ready for the great biomyths, visceral legends.

Imagine this game. Groups of people with highly constrained artificial behaviours moving through zones with different functions (like: magic, camouflage, enlargement, reversal, disparity). Gives you: zone-shifts, time-shifts, identity-shifts. No light-pen needed to work out that potential.

Dare we talk about art and social modelling?

We are very much concerned with generating futuribles – maybe that's because the more we can dream up alternative futures the more changeable the present can become. And change is what we are all about – change for its own sake. That is the essence of behaviourist art, and generating change is the aim of the behaviourist artist.

We could talk about the levels of resolution for examining two classes of art system – the discrete and the continuous. That's like classical and behaviourist art.

How about the notion of secret reciprocity?

Cybernetics will have come of age when we no longer notice the hardware, where the interface is minimal. Same goes for art?

The cybernetic age is an age of silences. Same goes for music?

Artist on the campus. We can create new rituals in the centres of learning. We can introduce art as visual matrix for the varied discourse of a university. To hell with commissioned monuments!

Is it useful to discuss the thermodynamics of an artwork? An artwork is hot when it is densely stacked with information bits, highly organised and rigidly determined. Hot artwork admits of very little feedback in the system artifact/observer, its really a one way channel; pushing a message from the artist, out through the artwork into the spectator.

Call it cool when the information bits are loosely stacked, of uncertain order, not clearly connected, ambiguous, entropic. Then the system allows the observer to participate projecting his own sense of order or significance into the work, or setting up resonances by quite unpredicted interaction with it. We must also consider the cut-out mechanism which operates when an artwork overheats; when it is too hot, too densely stacked with an overburdened accumulation of bits, a sort of infinitely inclusive field. Then the system switches to a very cool state and feedback of a high order is possible.

Behaviourist art has two principle aspects – the biological and the social. it will be more or less visceral, more or less groupy.

Great art sets up systems of attitudes which can bring about the necessary imbalance and dispersal in society whilst maintaining cultural cohesion. For a culture to survive it needs internal acrimony (irritation), reciprocity (feedbacks), and variety (change). Enter art.

With heart-swopping behind us, what about behaviour – transplants?

The process structuring of artworks must inevitably reflect the substructure of behaviours in our cybernated ecology. Gives you: video analogues of processes which may trigger new behaviours.

Art now comes out of a passionate affair with the future. Let's take into account ESP, astrology, divination by tarot, the whole psychic scene, and work out scenarios for the astral plane. Let the mediums give the message. Remember! Black and White magic is easily reproduced.

If we are to keep art schools let them be structured as homeostatic organisms, living, adaptive instruments for generating creative thought and action. But first – more artists and scholars – fewer clerks and boy scouts. no more phoney – liberal blind man's bluff. Within a behaviourist framework the creative interplay of reason, passion and chance can take place.

The CAM – concept is essentially a futurible – anticipatory and speculative, depending for its viability on an understanding of the past. As a projection of our behaviour – based culture it is intended to be a scenario which is neither surprise – free or definitive. It is an Alternative. The idea of the alternative or multifold alternatives is becoming the very core of art as it progresses. As in science and sociology, to which it aspires from time to time to relate, generating alternatives futures seems to be essential to the internal development of art.

Art creates mythic futures. The mythology of change and uncertainty and the ritualisation of the will to form combine in the behaviourist art. "Only through myth and the structures it requires can we combine the necessary paradox of definition and ambiguity, of order and uncertainty, of the tangible and the infinite". Levi-Strauss.

In the post-industrial society is not technology that will carry us through so much as psychotechnology. That may take us beyond Skinners behavioural engineering into the shadow lands, the futuribles, the speculative, astrological, dreamed-up, out-of-body, future behaviours. We may not have reached the frontiers of parapsychology, but when we do- wham ! Instant communications with no media. Total telepathy, waves of alternative behaviours surging on from creative impulses of the mind. A hardline software culture, always being rerouted, conditioned only to branch.

Art is now a form of behaviour

Message ends.

Published in: 1970, Control, N° 5, London, p.3
Stiles K, Selz P (eds) (1996) Theories and documents of contemporary art, University of California Press, pp 396, 489–91

References

1 Maturana H, Varela FJ (1987) The tree of knowledge: the biological roots of human understanding. New Science Library, Boston
2 Ascott R (1995) The architecture of cyberception. In: Toy M (ed) Architects in cyberspace. Academy Editions, London, pp 38–41
3 Ascott R (1966) Behaviourist art and the cybernetic vision. Cybernetica, Journal of the International Association for Cybernetics (Namur) 9: 247–264, 10: 25–5
4 Ascott R (1964) The construction of change. Cambridge Opinion, Cambridge, pp 37–42
5 Ascott R (1983) La Plissure du Texte. In: Popper F (ed) Electra. Musée d'Art Moderne, Paris, pp 398–399
6 Ascott R (1989) Gesamtdatenwerk. Konnektivität, Transformation und Transzendenz. Kunstforum (Köln), September/Oktober, 103: 100–109.
7 Ascott R (1993) Telenoia. In: Adrian R (ed) On Line – Kunst im Netz. Steirischen Kulturinitiative, Graz, pp 135–146

8 Ascott R (1996) Identity in cyberspace: pilot project for a european cyberspace collegium. Commission of the European Communities, D.G.XXII, Brussels. (CD ROM)
9 Here I follow David Chalmers' argument for the "irreducibility" of consciousness. See: Chalmers D (1996) The conscious mind: in search of a fundamental theory. Oxford University Press, New York
10 Cyberpsychic Nodes, Parque do Xingu, Brazil, May 1997. Expedition team: Ascott R, Domingues D, Fraga T, Prado G
11 du Noüy L (1936) Biological time. Methuen, London, p 110
12 Ascott R (1996) Is there love in the telematic embrace? Behaviourables and futuribles. In: Stiles K, Selz P (eds) (1996) Theories and documents of contemporary art. University of California Press, Berkeley, pp 396, 489–498
13 Bedworth J, Davies C, Harrison D, Hunt G, Landa K, Nechvatal J, Rogala M, Scott J, Seaman B, Vesna V
14 URL: http://caiiamind.nsad.newport.ac.uk
15 Ascott R (1996) The planetary collegium: art and education in the post-biological era, Artlink (Sydney).16 (2 & 3): 51–54
16 Ascott R (1993) From appearance to apparition: communications and culture in the cybersphere. In: Verostko R (ed) FISEA, Fourth International Symposium on Electronic Arts. MCAD, Minneapolis, pp 5–12

Biographies

Prof. Roy Ascott

Centre for Advanced Inquiry in the Interactive Arts (CAiiA),
University of Wales College, Newport

A pioneer of cybernetics and telematics in art, Ascott has been creating major global networking projects since 1980 including "la Plissure du Texte: a Planetary Fairytale" for Electra at the Museum of Modern Art, Paris, "Laboratory Ubiqua" for the Venice Biennale, "Aspects of Gaia" for Ars Electronica, Linz and "Telenoia" for the V2 Organisation, Holland. He first established the role of cybernetics in art in the 1960s with his "Change-paintings" and such seminal texts as "Behaviorist Art and the Cybernetic Vision". The relationship between consciousness and technology, between artificial life and emergent mind, between shamanism and telematics, are at the centre of his current projects, for which he lived with the Kuikuru Indians in the Mato Grosso in May 1997.
He has been International Commissioner for the Venice Biennale, and a jury member of Prix Ars Electronica Linz, and of the Interactive Media Festival San Francisco, and is currently in the jury chamber of the NTT InterCommunication Center Biennale, Tokyo. He is a member of the Arts Council of England.
A widely published theorist, his texts are translated into many languages. He is the founder-director of the Centre for Advanced Inquiry in the Interactive Arts (CAiiA) at the University of Wales College, Newport where he supervises, largely online, a research community of leading media artists based throughout the world.

Prof. Louis Bec

Institut Scientifique de Recherche Paranaturaliste
Inspecteur de la Création Artistique, Ministère de la Culture, Sorgues

Louis Bec lives and works in Sorgues (France Provence). His work perfidiously tackles with the relationships between arts, sciences and technologies.
The only living zoosystemician, he deals with a fabulatory epistemology based on Artificial Life and Technozoosemiotic (methodology, modelisation, systemic, taxonomy, ethology, bio-informatic ...).
In 1972, he founded the Scientific Institut of Paranaturalistic Research (ISRP) which allowed him to develop his researches on heuristic systems: UPOKRINOMENES and UPOKRINOMENOLOGY.
Various exhibitions and events, seminars and lectures:
- Fondation du Futur (Nicolas Ledoux) Arc et Senans, (France)
- Biennale de Sao Paulo, (Brasil)
- Frasso Telessino, (Italy)
- PhotoVISION, Hannover (Germany), Vienna (Austria), Zürich (Switzerland)
- Musée National d'Histoire Naturelle, (Paris)
- Theorie du Chaos et Art, Graz (Austria)
- Artificial Life II, Santa Fe (USA)
- IMAGINA, Monaco (Monaco)
- Revue Virtuelle, Centre Georges Pompidou, Paris (France)

Organization of a number of important events:
„LE VIVANT ET L'ARTIFICIEL", Festival d'Avignon, 1984
ART/COGNITION 92, Aix en Provence, 1996/97
EMAITRE, project of telerobotic with partnerships in Russia and Italy
Personal works:
Researches: collaboration about animal behavior and artificial life with the French National Center for Scientific Research (CNRS) and several other laboratories: Strasbourg University Photonic Laboratory (ENSPS/ARTCAP), Institut Non-Linear (INL) Sophia Antipolis Nice, Behavioural Biology Laboratory University of Grenoble, CYPRES Aix en Provence ...

Prof. Donna J. Cox

University of Illinois at Urbana-Champaign (UIUC), National Center for Supercomputing Applications (NCSA), Urbana-Champaign, IL, USA

January 1993- Professor, School of Art & Design,
University of Illinois @Urbana-Champaign (UIUC)
March 1992-August 93 Co-Director, Scientific Communications and Media Systems (SCMS),
National Center for Supercomputing Applications (NCSA)
August 1990- Associate Director for Technologies, School of Art, UIUC
August 1990-January 1993 Associate Professor, School of Art, UIUC
August 1989 -March 1992 Associate Director for Numerical Laboratory Programs, NCSA
January 1989-August 1990 Project Leader and Principal Investigator,Renaissance Experimental Laboratory
Awards and Achievements:
* 1992 Publisher's Award, computer animation, from Fresh Electronic Publishing. Cash award.
* National Computer Graphics Association (NCGA) Animation Competition, 2nd place Academic Category, 1992. Anaheim, March 9-12,1992, Video Theatre

Donna J. Cox has exhibited computer images and animations in more than 100 invitational and Juried exhibits in the past nine years, including shows at the Bronx Museum of Art in New York , Everson Art Museum in New York, Feature Gallery in Chicago, Feature Gallery in Soho New York City, Fermilab in Chicago, Museum of Contemporary Photography in Chicago, Frick Art Museum in Ohio, Oregon Art Museum and Milwaukee Art Museum in Wisconsin.

Monika Fleischmann

GMD German National Research Center for Information Technology, Bonn, Germany
Monika Fleischmann (1950) is a researcher and artist. Since 1992 she has been head of the media culture projects at the Institute for Media Communication at the GMD German National Research Center for Information Technology in Sankt Augustin near

Bonn. Pivotal to her artistic and scientific work is the rediscovery of the perception of the human senses using a virtual environment. Fleischmann studied the educational theory of art, play and theatre at the HdK Berlin and fashion and costume design in Zürich and has worked as a teacher, author, dramatic adviser and designer. In 1988 she was co-founder of ART + COM in Berlin, a research institute for computer-based design and representation.

Dr. Cynthia Goodman

Cynthia Goodman is a curator and producer of interactive multimedia art. Former Director of the IBM Gallery of Science and Art, New York, where she organized the landmark Computers and Art exhibition, her accompanying publication, Digital Visions: Computers and Art, serves as a textbook in the field. Most recently, Goodman was Co-Director with Nam June Paik of the InfoART Pavilion at the Kwangju Biennale in Korea, an international exhibition that showcased the top artists in the multimedia art field. Her most recent publication CD-ROM, InfoART: The Digital Frontier from Video to Virtual Reality, produced at Rutt Video Interactive, NY. Also active in museum automation, she participated in the John Paul Getty Trust Museum Prototype Project and was based at the Solomon R. Guggenheim Museum, New York.
Goodman has acted as advisor to numerous corporations including IBM, Polaroid, and Time Warner. Appointed Fellow at the Center for Advanced Visual Studies, Massachusetts Institute of Technology, she was Director of Arttransition '90, an international conference on art, science and technology. She was also one of the Directors of Artec '91 in Japan, the first international biennial of art and technology. She has served as Juror for the annual SIGGRAPH art show and was recently on the Advisory Committee for Women and the Art of Multimedia, an international conference organized by the National Museum of Women in the Arts, Washington, D.C.
She has organized and installed exhibitions for numerous institutions including The Metropolitan Museum of Art, New York; the Whitney Museum of American Art, New York; the IBM Gallery of Science and Art; the Contemporary Arts Center, Cincinnati, OH; the Everson Museum of Art, Syracuse, NY; the National Building Museum, Washington, D.C.; and the Centre Georges Pompidou, Paris, France.

Prof. Erkki Huhtamo

University of Lapland, Finland

Erkki Huhtamo is media scholar, writer and curator. He is currently professor of media studies, University of Lapland, Finland. Professor Huhtamo's specialities include media history and the aesthetics & development of media art.
Mr Huhtamo has published numerous studies and articles (in ten languages), among them "Encapsulated Bodies in Motion: Simulators and the Quest for Total Immersion", in Critical Issues in Electronic Media, edited by Simon Penny (New York: SUNY Press, 1995), "Seeking Deeper Contact. Interactive Art as Metacommentary", Convergence, Vol.1, N:o 2 (Autumn 1995), pp. 81-104 (University of Luton & John Libbey,

U.K.), "Time Machines in the Gallery. An Archeological Approach in Media Art", in Immersed in Technology. Art and Virtual Environments, edited by Mary Anne Moser & Douglas McLeod, Boston: The MIT Press, 1996 (forthcoming), "Digitalian Treasures, or Glimpses of Art on the CD-ROM Frontier", in Electronic Art in the 90's, edited by Lynn Hershman, Seattle: The Bay Press (forthcoming).
Mr Huhtamo was contributing editor of InterCommunication, N:o 14 ("An Archeology of Moving Image Media", Autumn 1995), Tokyo: ICC. His most recent book is titled "The Archeology of Virtuality" (in Finnish, 1995).
Mr Huhtamo has lectured widely in Europe, the United States, Canada, Australia and Japan. He has curated several international exbitions of media art, including ISEA 94 Exhibition (the Museum of Contemporary Art, Helsinki, 1994, with Asko Makela), Toshio Iwai Retrospective (ZKM, Karlsruhe & OTSO Gallery, Espoo, 1994, with Paivi Talasmaa & H-P Schwartz) and Digital Mediations (Art Center College of Design Gallery, Pasadena, Ca.,1995, with Stephen Nowlin). Mr.Huhtamo has also written and directed three television series on media culture for YLE (the Finnish Broadcasting Company). The most recent is titled "The Archeology of the Moving Image" (under production).

Prof. Toshiharu Itoh

Tama Art University, Tokyo, Japan

Itoh Toshiharu was born in 1953. He is an art critic specializing in photographic and other image media.
He is committee member of ICC InterCommunication Center Japan.
Mr. Itoh's critical work spans from the history of art, photography and media theory, to design, music, film and 19th and 20th century culture.
A highly prolific author, he in 1987 "Georama Theory" won the Suntory prize for literature. Presently researching museums, libraries and other cultural facilities, his recent work has offered fresh new insights into architecture and urban design theory. Some of Prof. Itoh's published works include "The Photographic City" (Shashin Toshi), "On the Body in Ruins"(Seitai Haikyo Ron), "Georama Theory", "Discommunication", "Magical Hain", "The Machine Art".

Dr. Michael Klein

Born 1960 in Wuppertal, Germany, studied physics and philosophy at the University of Wuppertal, Germany, between 1980 and 1983, studied physics at the University of Tübingen, Germany, between 1983 to 1989, worked as assistent (-professor) at the Institute for Physical and Theoretical Chemistry at the University of Tübingen, Germany, between 1988 and 1992, completed his doctor thesis in 1992 about chaos theory and complex systems at the department of theoretical chemistry at the University of Tübingen with O.E. Rossler, worked as assistent professor in science and arts with Peter Weibel at the Städelschule - Institute for New Media, Frankfurt Germany, between 1992 and 1994, since September 1994 he is director of the refounded INM-Institute for New Media, Frankfurt, Germany, founded the INM-Numerical Magic Gesellschaft für neue Medien mbh in May 1996.

Wrote numerous scientific publications in the fields of nonlinear dynamics, chaos and theory of complexity as well as interactive arts and sciences.
Dr. Klein is co-editor of the book "A Chaotic Hierarchy", Gerold Baier and Michael Klein, World Scientific Singapore 1991; he is co-publisher of the book "Profiles, INM - Institut für Neue Medien", F.A. Bechtoldt and Michael Klein, INM Institut für Neue Medien publishers 1996, he founded an interdisciplinary research group ENGADYN in 1990.

Prof. Machiko Kusahara

Tokyo Institute of Polytechnics, Media Art at the Faculty of Arts, Tokyo, Japan

Machiko Kusahara is an Associate Professor of Media Art at the Faculty of Arts, Tokyo Institute of Polytechnics. She has been teaching computer graphics and media since 1986 and is a prolific writer on the subject of new media since 1984. Her writing is published in many Japanese publications, including the widely read Asahi Newspaper, as well as InterCommunication and SuperDesigning magazines. She has published "Computer Graphics Anthology" (1989, Bunkensha Publishing Co.) and "Computer Graphics Access" (1992, Bunkensha Publishing Co.) which altogether cover the history of computer graphics with seventeen laserdiscs and two textbooks. She co-authored many publications including "Hyper Image Museum" (CD-ROM, 1994, F2 Publishing), "Aesthetic of A-Life" (Yosensha, 1994), and "Techno-culture Matrix" (NTT Publishing, 1994). In addition to these professional commitments, Kusahara serves as a member of the Program Committee of NICOGRAPH (Nippon Association of Computer Graphics), the Planning Committee for MultiMedia Grand Prize of the MultiMedia Association of Japan, and a co-founder and a Program Committee member of Digital Image, the largest group of digital artists and designers in Japan. She is also a member of the ACM/SIGGRAPH, Information Processing Society of Japan, Japan Society of Image Arts and Science, NICOGRAPH, 3D FORUM and A-Life Network of Japan. Kusahara's career began in 1983 with the management of the SIGGRAPH Tokyo office. The SIGGRAPH Traveling Art Show in 1985 marked her debut as a curator in media art. Since then she has been commissioned for planning and exhibitions of World Design Exposition (1989), Metropolitan Museum of Photography (1988-95), Sony ArtArt Exhibitions (1990-91), NTT InterCommunication Center (1990-), among others.

Maria Grazia Mattei

Milano, Italy

Maria Grazia Mattei is a journalist and expert in new technologies of communication, and the new media consultant for Camera di Commercio di Milano. She has organized and covered numerous multimedia, communications and cultural events and conferences from the around the world.
For many years she has been cooperating with cultural Institutions (Triennale of Milan,

Palazzo Fortuny in Venice, Biennale of Venice), specialized magazines (Virtual, Gulliver) as well as the Italian newspaper "Il Sole 24 Ore".
She curated an exibition consacrated to the new media "Oltre il Villaggio Globale" ("Beyond the global village") for the Triennale of Milan (on April 27-June 27, 1995). She has concieved and personally followed exhibitions, international reviews and conferences on the relationship among art, technology and communication, such as "Arte e Nuove Tecnologie" (Pavia, 1984), "Mondi Virtuali" (Venezia, 1990), "Computer and Art" (Lugano, 1991), "Pros and Cons the New Technologies in the Audiovisual Field" (La Biennale of Venice, 1994).
She has been in international juries (Interactive Multimedia Festival, Los Angeles 1994, IDMA, Toronto 1995-96-97; Milia d'Or, Cannes 1997) and since 1990 she has been cooperating with Imagina in Montecarlo. She is a member of the International Committe of Siggraph '96 in USA.
Since 1989, she is the director of Mediatech, Forum of IBTS (International Broadcasting Telecommunication Show) on new technologies of communication. She has curated for Mediatech international lectures, panels, seminars, conferences on evolving scenarios of new medias and broadcasting.

Dr. Gottfried Mayer-Kress

College of Health and Human Development, Penn State University

Center for Complex Systems Research, Beckman Institute, Department of Physics, University of Illinois at Urbana-Champaign, USA

Gottfried Mayer-Kress received his Diploma in Theoretical High Energy Physics from the University of Hamburg, Germany in 1978 and his Dr. rer. nat. from the Institute of Theoretical Physics and Synergetics, University of Stuttgart, Germany in 1984. From 1984 to 1987 he was director's postdoc at the Center for Nonlinear Studies, Los Alamos National Laboratory, Los Alamos, NM. From 1987 to 1988 he held research appointments at the Laboratory for Biological Dynamics and Theoretical Medicine, University of California at San Diego and at the Crump Institute for Medical Engineering at the University of California at Los Angeles. From 1988 to 1991 he was Visiting Assistant Professor at the Mathematics Department of the University of California at Santa Cruz Since 1992 he is Visiting Assistant Professor at the Center for Complex Systems Research and Department of Physics at the University of Illinois at Urbana Champaign. He also held research appointments at the Department of Chemical Engineering of Princeton University, at the Observatoire de Nice, France and at the Santa Fe Institute. His research interests cover many areas of nonlinear dynamics and their applications.
Research Interests:
Non-linear dynamics, stochastic processes and complex adaptive systems and their applications to problems in physics, medicine, crisis management. Data analysis and modelling with nonlinear methods. Control of chaotic systems, scientific visualization/audification of complex systems and their cognitive representations. Interdisciplinary research on distributed information and simulation systems, Internet, and World Wide Web (www).

Laurent Mignonneau

ATR Advanced Telecommunications Research Laboratories in Kyoto
IAMAS International Academy of Media Arts and Sciences, Gifu Japan

Laurent Mignonneau is Artistic Director and Researcher and at the ATR Advanced Telecommunications Research Laboratories in Kyoto, Japan and Artist-in-Residence and Lecturer at the IAMAS International Academy of Media Arts and Sciences in Gifu Japan.
Laurent Mignonneau studied Fine Arts, Video, Performance, Music and Electronics at the Academy of Fine Arts in Angouleme France. In 1992 he received a European Grant to study at the post graduate Institute for New Media Frankfurt in Frankfurt Germany. In 1992 Mignonneau started collaboration with the Austrian Artists Christa Sommerer.
Their works have been shown world wide in all important international exhibition of interactive arts, for example at the Wilhelm Lehmbruck Museum Duisburg, the Kunsthalle Bonn, the Biennale de Lyon 95/96, the Henie Onstad Kunstcenter Oslo, the Center for the Arts - Yerba Buena Gardens San Francisco, the Neue Galerie Graz, the "Kwangju Biennale '95" in Korea, the "Triennale di Milano" in Milano Italy, the Nagoya City Art Museum Japan and the Museum of Modern Art in Helsinki.
Mignonneau and Sommerer have lectured world wide at international symposia and conferences on art, interactivity, artificial life and communication art. Their scientific work has been published in major publications, such as for example in "Artificial Life V" by Langton, Christopher (ed.) MIT Press, in the "Complexity Journal", Wiley & Sons, September 97 and at the Siggraph Visual Proceedings 93, 95 and 96, ACM Siggraph New York .
In 1995 Mignonneau and Sommerer organized and chaired an international symposium on Art and Science, the "ART-Science-ATR" symposium at the ATR Advanced Telecommunications Research Laboratories in Kyoto, which laid the basis for their publication of "Art @ Science" at Springer-Verlag Wien New York.

Michael Naimark

Interval Research Corporation, Palo Alto, USA

Michael Naimark spent twelve years as an independent media artist before joining Interval Research Corporation in 1992. He was instrumental in making the first interactive laserdiscs in the late 1970s at MIT and has worked extensively with projection and immersive virtual environments. He has consulted on new media for various institutions and his artwork has been exhibited internationally.
Naimark has held faculty appointments at the San Francisco Art Institute, San Francisco State University, California Institute of the Arts, M.I.T., the University of Michigan, and is on the Editorial Boards of Presence and Leonardo Electronic Almanac. He created a B.S. in Cybernetic Systems as an independent major from the University of Michigan in 1974 and received an M.S. in Visual Studies and Environmental Art from M.I.T. in 1979.
Selected Art Exhibitions:

1995 Center for the Arts, San Francisco 1995 Interaction '95, Gifu, Japan 1995 University Art Museum, Berkeley 1994 ISEA, Helsinki 1993 Banff Centre for the Arts, Canada 1992 SIGCHI "Future Scenario" video program, Monterey, CA 1992 Images du Futur, Montreal 1992 La Triennalle di Milano, Milan 1991 SIGGRAPH Tomorrow's Realities Gallery, Las Vegas 1991 Multimediale, Karlsruhe 1989 Kanagawa International Art and Science Exhibition, Tokyo 1989 World Financial Center, New York 1988 Kennedy Center for the Performing Arts, Washington, D.C. 1987-- Exploratorium, San Francisco 1986-88 Madeleine Metro Station, Paris 1984 San Francisco Museum of Modern Art 1980 Center for Advanced Visual Studies, M.I.T.
Selected Research Consultations:
1991-92 M.I.T. Media Lab "Movies of the Future" Program 1987-90 Lucasfilm Learning and Games Divisions 1987-90 Apple Multimedia Lab and Human Interface Group 1987 Mattel 1986 NY Museum of Modern Art 1986 Microsoft 1985 Panavision 1982-84 Atari Research Lab 1982 WED Enterprises (Disney Research) 1981 National Geographic Society Office of the President 1979-81 M.I.T. Arts and Media Technology Facility (Media Lab).

Dr. Ryohei Nakatsu

Ryohei Nakatsu received his B.S., M.S. and Ph.D. degrees in electronic engineering from Kyoto University in 1969, 1971 and 1982, respectively. After joining NTT in 1971, he mainly worked on speech recognition technology. Since 1994, he has been with ATR (Advanced Telecommunications Research Institute) and currently is the president of the ATR Media Integration & Communications Research Laboratories. His research interests include emotion extraction from speech and facial images, emotion recognition, nonverbal communications, and integration of multi-modalities in communications.
He is a member of the IEEE, the Institute of Electronics, Information and Communication Engineers Japan (IEICE-J), as well as the Acoustical Society of Japan.

Prof. Przemyslaw Prusinkiewicz

Department of Computer Science, The University of Calgary, Canada

Przemyslaw Prusinkewicz is a Professor of Computer Science at the University of Calgary. He has been conducting research in computer graphics since the late 1970s. In 1985, he originated a method for visualizing the structure and the development of plants based on L-systems, a mathematical model of development. He is a co-author of three textbooks and two monographs, "Lindenmayer Systems, Fractals and Plants" (Springer-Verlag 1989) and "The Algorithmic Beauty of Plants" (Springer-Verlag 1990), as well as approximately 50 technical papers. His current research includes the mathematical modeling and visualization of various aspects of morphogenesis.
Professor Prusinkiewicz holds an M.S. and Ph.D., both in Computer Science, from the Technical University of Warsaw. Before joining the faculty of the University of Calgary, he was Professor at the University of Regina, and Assistant Professor at the Univer-

sity of Science and Technology of Algiers. He was also a Visiting Professor at Yale University (1988), at L'Ecole Polytechnique Federale de Lausanne (1990), and an invited researcher at the University of Bremen (1989) and the Centre for Tropical Pest Management in Brisbane (1993, 1994).

Philippe Quéau

UNESCO, Paris
IMAGINA, Monte Carlo

Philippe Quéau is the Director of the Information and Informatics Division at UNESCO. He formerly was the IMAGINA Program Chairman and INA Director of research. A graduate of the Ecole Polytechnique in Paris and a graduate engineer of the Ecole Nationale Superieure des Telecommunications, Philippe Quéau was Director of Research at the Institut National de l'Audiovisuel (INA), a public organisation under the French Ministry of Communication. Philippe Quéau is a member of the Ministerial Research Commitee reporting to the French Ministry of Culture and Communication. He is participating in various study groups working for the French government in the field of telecommunications and new imaging techniques. He is a consultant for the Commission of the European Union and for the UNESCO.
In 1980 he set up the Image Research Group at INA. In 1981 he was instrumental in creating IMAGINA, the Monte Carlo International Forum on new Images, for which he acted as programme organiser ever since. IMAGINA is the leading European event devoted to computer imaging, virtual reality and cyberspace. More than 6000 delegates from 30 countries attended Imagina 96.
Philippe Quéau has written three books of aesthetic and philosophical analysis of developments in computer graphics, digital imaging and virtual reality techniques applied to audiovisual and artistic creation :

* Eloge de la Simulation - De la vie des langages a la synthese des images - published by Champ Vallon/INA, 1986 (publication of Japanese translation by Tokyo Shoseki in preparation).
* METAXU : Theorie de l'Art Intermediaire - published by Champ Vallon/INA - 1989.
* Le Virtuel : Vertus et Vertiges - published by Champ Vallon/INA - 1993.

Dr. Thomas S. Ray

ATR Human Information Processing Research Labs, Kyoto, Japan

Thomas Ray earned undergraduate degrees in biology and chemistry at Florida State University. He received his Masters and Doctorate in Biology from Harvard University, specializing in plant ecology. He was a member of the Society of Fellows of the University of Michigan at Ann Arbor. He joined the faculty of the University of Delaware, School of Life and Health Sciences in 1981, where he is now an Associate Professor. In 1993, he received a joint appointment in Computer and Information Science at U. of Delaware, and was appointed to the External Faculty of the Santa Fe

Institute. In August of 1993, he joined the new Evolutionary Systems Department at ATR (Advanced Telecommunications Research International) Human Information Processing Research Labs in Japan, as an invited researcher.
Dr. Ray is a tropical biologist who since 1974 has studied the evolution and ecology of a variety of organisms inhabiting rain forests. His work has focused primarily on the foraging behavior of vines in the family Araceae, however, he has also studied ants, butterflies and beetles. Most of his field work has been conducted in Costa Rica. Since 1982, he has worked principally at Finca El Bejuco biological station located in the lowland rain forests of northern Costa Rica, which he built, and owns and operates. He is actively engaged in rain forest conservation in Costa Rica.

Peter Richards

The Exploratorium, San Francisco, USA

Peter Richards is Director of Arts Programs at The Exploratorium, a museum of science, art and human perception located in San Francisco. He is responsible for commissioning artists to create works that are related thematically with exhibits and programs of the museum. He joined the museum staff in 1971 and has concurrently been creating artworks in public places that explore the relationship between people, places and the environment.
Mr. Richards received his B. A. degree of sculpture from Collorado College in 1966 and his M. F. A. from the Rinehart School of Sculpture in Baltimore in 1969. He has lectured at the Center for Experimental and Interdisciplinary Arts at San Francisco State University, the Ecole d'Art Aix en Provence, and San Francisco State Art Institute and has organized seminars and workshops on the interrelatedness of art and science. He has consulted for numerous science museums both in the States and abroad.
His most recent artwork, "Clapotis", a wave activated sound sculpture, was installed on Lake Geneva, Switzerland in June, 1996. He has permanent outdoor works installed at Artpark, Lewiston, New York, Olympia, Washington, Palo Alto, California and San Francisco. He has received fellowships from the Californian Arts Council and the Fleishhaker Foundation. The design for Byxbee Landfill Park, done in collaboration with Michael Oppenheimer, and George Hargreaves Associates received the American Landscape Society Award in 1993.

Prof. Otto E. Rossler

Division of Theoretical Chemistry, University of Tübingen, Germany

Otto E. Rossler, born as an Austrian in Berlin, finished his formal education with an immunological dissertation in 1966. In 1969, he won a competitive visiting appointment offered by the Center for Theoretical Biology at the State University of New York at Buffalo, with a manuscript on biogenesis. Art Winfree introduced him to chaos in 1975. In 1976 he obtained a tenured faculty position in Theoretical Biochemistry at the University of Tubingen. He claims the whisper in a glass of mineral water is hyperchaos. He also believes in an equation for a brain (1974), meaningful computer faces (1986) and the endo approach to physics (unless falsified). Most recently, he proposed Lampsacus, hometown of mankind on the Internet (http://www.cs.wayne.edu/~kjz/lampsacus/).

Prof. Itsuo Sakane

IAMAS International Academy of Media Arts and Sciences, Gifu, Japan

Itsuo Sakane is the President of IAMAS International Academy of Media Arts and Sciences in Ogaki, Japan. He was Professor at the Faculty of Environmental Information at Shonan Fujisawa Campus of Keio University (since 1990). He has been teaching an Introductory course to the Science-Art and other related courses including the Multimedia Seminar. As a former senior staff writer for the Asahi Shimbun he had covered the border field between art, science and technology for more than 30 years.
Selected Publications of books; "The Coordinate of Beauty" (1973), "Katachi Mandala--Thinking through Seeing" (1976), "The Museum of Fun Book I, II"(1977, 1982). "Trip to the World in the Border-area between Science and Art"(1985), "Essays on Science and Art" (1986). Selected Translation of books; "Magic Mirror of M.C.Escher", "Escher on Escher", "M.C.Escher: His Life and Complete Graphic Works", etc.
Selected Articles and papers in journals; 'Art et technologie-a la recherche d'un nouveau rapport' (ART PRESS H.S. No.12, 1991 p.53-58), 'Recovering The Wholeness of Art: Information Versus Material' (LEONARDO, vol. 24, No.3. p.259-261,1991, Pergamon Press), 'Recovery of the Sense of Body', (1991.11.13. Culture page, Asahi Shimbun)
Selected Exhibitions; He worked as a Director and Curator; Phenomenart" (1989), "Wonderland of Science-Art: Invitation to the Interactive-art"(1989). " Science-Art Gallery" at Japan Pavilion in Expo 92 at Seville, Spain, "Interaction95" and "Interaction97"(Ogaki), "Art and Science" exhibitions at the NTT InterCommunication Center (ICC) in Tokyo 97.
A commissioner for Japanese Artists for ELECTRA Exhibition in Paris(1982), etc.
Award: Japan Culture & Design Award (1982).
International Co-editor for LEONARDO, a Journal of ISAST (International Society for Arts, Sciences, and Technology). Born in Tsingtao, China in 1930. Graduated from Tokyo University (Department of Architecture). Nieman Fellow at Harvard University (1970-71).

Prof. Hans-Peter Schwarz

Media Museum, ZKM Karlsruhe, Germany

Hans-Peter Schwarz studied Visual Communication at the Fachhochschule Bielefeld and Art History, Modern German Literature and European Ethnology at Marburg University, where in 1982 he obtained a doctorate with a thesis on artists houses in the area of tension between court and city.

From 1983 to 1990 he was custodian at the Deutsches Architekturmuseum (DAM) in Frankfurt, responsible primarily for exhibitions and the permanent collection. In his function as Deputy Museum Director he represented the DAM academically at an international level.
He is author of numerous publications on architecture and art history. He also taught at the Universities of Marburg, Trier and Frankfurt/Main and at the Fachhochschule Darmstadt.
In 1992 he was appointed Director of the Media Museum at the ZKM Karlsruhe (Center for Art and Media Technology) – Zentrum fuer Medien Kunst und Technologie) in Karlsruhe.
Since 1994 he also holds a Professor position for Art History at the Hochschule fuer Gestaltung in Karlsruhe.

Prof. Jeffrey Shaw

ZKM Karlsruhe, Germany

1963 Architecture at the University of Melbourne
1964 Art History at the University of Melbourne
1965 Sculpture at the Brera Academy of Art, Milan
1966 Sculpture at St Martins School of Art, London
1970 - 1980 Founding member of The Eventstructure Research Group
1989 Guest professor at the Academy of Art, Rotterdam
1990 Guest professor at the Rietveld Academy, Amsterdam
Since 1991 Director of the Institut fuer Bildmedien, at the ZKM / Zentrum fuer Kunst und Medientechnologie Karlsruhe.
Since 1995 Professor at the Hochschule fuer Gestaltung Karlsruhe
Prizes:
1990 L'Immagine Elettronica, Ferrara, Italy
1990 Ars Electronica, Linz, Austria
Selected Works 1967-1995:
The Legible City (with Dirk Groeneveld), Bonnefanten Museum, Maastricht 1988
The Imaginary Museum of Revolutions (with Tjebbe van Tijen)
Brucknerhaus, Linz 1989
Alice's Rooms 'International Art & Science Exhibition', Kawasaki 1989
Revolution 'Imago', KunstRai, Amsterdam 1990
The Virtual Museum 'Das Belebte Bild', Art Frankfurt, Frankfurt 1991
Disappearance 'The Binary Era', Musee d'Ixelles, Bruxelles 1992
Eve 'MultiMediale 3', ZKM Center for Art and Media, Karlsruhe 1994
The Golden Calf 'Ars Electronica', Linz 1994
Place - a Users Manual, Neue Galerie, Graz 1995

Since the late 60's Jeffrey Shaw has pioneered the use of interactivity and virtuality in his many art installations. His works have been exhibited worldwide at major museums and festivals. Probably best known is his LEGIBLE CITY (1989) where the spectator can bicycle in a computer generated virtual urban landscape of words and sentences. For many years he was living in Amsterdam where he cofounded the Even-structure Research Group (1969-80). At present Shaw is director of the Institute for Visual Media at the ZKM Center for Art and Media Karlsruhe, Germany He leads a unique research and production facility where artists and scientists are working together and developing profound artistic applications of the new media technologies.

Prof. Christa Sommerer

IAMAS International Academy of Media Arts and Sciences, Gifu Japan
ATR Advanced Telecommunications Research Laboratories, Kyoto, Japan

Christa Sommerer is Professor for Media Art at the IAMAS International Academy of Media Arts and Sciences in Gifu Japan. Since 1994 she additionally holds a position as Researcher and Artistic Director at the ATR Advanced Telecommunications Research Laboratories in Kyoto, Japan.
Christa Sommerer had originally studied Modern Sculpture at the Academy of Fine Arts Vienna as well as Botany at the University of Vienna Austria. In 1992 she became interested in new media and subsequently studied media art from 1992 to 1993 at the post graduate Institute for New Media Frankfurt in Frankfurt Germany. There, in 1992, she started collaborating with the French Artists Laurent Mignonneau on the creation of interactive computer installations, such as "Interactive Plant Growing", "A-Volve", "Phototropy", "Life Spacies" and "GENMA." From 1993 to 1994 Sommerer and Mignonneau worked as Artists-in-Residence at the NCSA National Center for Supercomputing Applications in Champaign Urbana, USA and in 1994 were invited as Artists-in-Residence by the ICC InterCommunication Center Tokyo.
Sommerer and Mignonneau's interactive arts works have been called "epoche making" (Toshiharu Itoh) as they pioneer the use of "natural interfaces" and a new language of interactivity combined with artificial life and communication. Sommerer and Mignonneau have won all mayor international awards for Interactive Art, such as the "Golden Nica" Award (1. prize) for Interactive Art at the Ars Electronica Festival in 1994 in Linz Austria, the "Ovation Award" at the Interactive Media Festival in Los Angeles in 1995 and the "MultiMedia Award '95" of the Multimedia Association in Japan.
Their interactive installations are considered as classics in the field of interactive media art and are subsequently included in mayor media art museums, such as the Museum of Contemporary Art Lyon, the Tokyo Metropolitan Museum of Photography, the Centre Georges Pompidou Paris, the Media Museum of the ZKM Karlsruhe, the Ars Electronica Museum and the NTT InterCommunication Museum Tokyo.

Prof. Wolfgang Strauss

Saar College of Fine Arts, Germany

Wolfgang Strauss is an architect and visiting professor in media design at the Saar College of Fine Arts. In 1988, he was co-founder of ART + COM in Berlin, concentrating primarily on virtual space and interface design. His work as a guest researcher at the GMD Institute for Media Communication in Sankt Augustin deals with the relationship between the human body and the digital image space. Strauss works with art and design students on experimental design methods for space-related installations and intermedia forms of representation.

Prof. Demetri Terzopoulos

Department of Computer Science, University of Toronto, Toronto, Canada

Demetri Terzopoulos is Professor of Computer Science and Electrical and Computer Engineering at the University of Toronto, where he leads the Visual Modeling Group, and is a Fellow of the Canadian Institute for Advanced Research. He received a PhD degree in Artificial Intelligence from MIT in 1984, and MEng and BEng degrees in Electrical Engineering from McGill University in 1978 and 1980, respectively.
From 1985-92 he was affiliated with Schlumberger, Inc., serving as Program Leader at research labs in Palo Alto, CA, and Austin, TX. During 1984-85 he was a research scientist at the MIT Artificial Intelligence Lab, Cambridge, MA.
His published works include more than 150 scientific articles, primarily in computer graphics and vision, but also in computer-aided design, medical imaging, artificial intelligence, and artificial life, including the recent edited volumes Real-Time Computer Vision (Cambridge Univ. Press '94) and Animation and Simulation (Springer-Verlag '95).
His research contributions have been recognized with several awards. In 1996 he was awarded the E.W.R. Steacie Memorial Fellowship by the Natural Sciences and Engineering Research Council of Canada.
He is a founding member of the editorial boards of the journals Medical Image Analysis, Graphical Models and Image Processing, and the Journal of Visualization and Computer Animation, and is a member of the IEEE, NYAS, and Sigma Xi.

Prof. Peter Weibel

Academy of Applied Arts, Vienna, Austria

Peter Weibel is Professor for Media Arts at the Academy of Applied Arts in Vienna, Austria.
For several years he headed the well known "Ars Electronica" Festival in Linz, Austria as its artistic director. He is also founder and former Director of the Institute for New Media, a well known Media Academy in Frankfurt Germany.
His numerous activities as curator includes his position as main organizer and curator of the 1995 and 1997 Biennale de Venice and his position as Director of the Museum of Modern Art in Graz, Austria.
Prof. Peter Weibel is Professor, Curator, Artist and Organizer, and has published several books and numerous articles on Media Art and Media Theory.

Bibliography

Literature:

Arnheim R (1969) Visual thinking. University of California Press, Berkeley

Ascott R (1991) Connectivity. Art and interactive telecommunication. Leonardo 24 (2)

Bachelard G (1985) The new scientific spirit. Beacon Press, Boston

Bachelard G (1934) Le nouvel esprit scientifique. Presses Universitaires de France, Paris

Bachelard G (1994) The poetics of space. Beacon Press, Boston

Barthes R (1978) Image, music, text. Farrar Straus & Giroux

Bateson G (1978) Mind and nature. Dutton, New York

Bateson G (1972) Steps to an ecology of mind. Ballantine, New York

Baudrillard J (1988) Selected writings. Poster M (ed) Stanford University Press, Stanford, CA

Baudrillard J (1996) The perfect crime. Verso Books

Bender G, Druckrey T (eds) (1994) Culture on the brink: ideologies of technology. Bay Press, Seattle, WA

Benedict M (ed) (1991) Cyberspace: first steps. MIT Press, Cambridge, MA

Benjamin W (1977) Das Kunstwerk im Zeitalter seiner technischen Reproduzierbarkeit. Suhrkamp Verlag, Frankfurt

Benjamin W (1978) Illuminations. Schocken Books, New York

Benthall J (1969–1979) Technology and art. Studio International. Monthly journal

Benthall J (1969–1979) Science and technology in art today. Thames and Hudson

Bishton D, Cameron A, Druckrey T (eds) (1991) Ten 8: Digital Dialogues. Vol. 2. Autumn. Birmingham: Ten 8 Ltd

Bohm D, Peat DF (1987) Science, order and creativity. Routledge

Bohm D (1980) Wholeness and the implicate order. Routledge

Brooks RA, Maes P (eds) (1994) Artificial Life IV : Proceedings of the Fourth International Workshop on the Synthesis and Simulation of Living Systems (Complex Adaptive Systems). MIT Press, Cambridge, MA

Burgin V. The End of Art Theory. Criticism and Postmodernity. Atlantic Highlands, Nj.: Humanities Press International Inc.

Cage J (1993) Silence. Lectures and writings. Middletown, Connecticut

Capra F (1982) The turning point. Bantam Books, New York

Capra F (1975) The tao of physics. Shambhala, Berkeley

Casti JL (1997) Would-be worlds. John Wiley and Sons, New York

Casti JL (1992) Paradigms lost. Abacus, London

D'Agostino P (1985) Transmission: theory and practice for a new television aesthetics. Tanam Press, New York

Davis D (1973) Art and the future. A sistory/prophecy of the sollaboration between science, technology and art. Thames and Hudson

Davis D (1977) Artculture. Harper and Row, New York

Dawkins R (1986) The blind watchmaker. WW Norton and Company, New York

Dawkins R (1976) The selfish gene. Oxford University Press, Oxford/New York

De Kerckhove D (1991) Communication arts for a new spatial sensibility. Leonardo 24 (2)

Dinkla S (1997) Pioniere Interaktiver Kunst. Edition ZKM: Cantz Verlag

Druckrey T (1993) Iterations. The new image. International Center for Photography, New York

Dunn D, Weibel P (eds) (1992) Eigenwelt der Apparate-Welt. Pioneers of Electronic Art. Ars Electronica, Linz

Emmer M (1996) The visual mind. Art and mathematics. MIT Press, Cambridge, MA

Escher MC (1971) The World of M. C. Escher. H. N. Abrams, New York

Felderer B (ed) (1996) Wunschmaschine Welterfindung. Springer, Wien New York

Felshin N (ed) (1995) But is it art? The spirit of art as activism. Bay Press, Seattle, WA

Feyerabend P (1993) Against method. Verso Books

Feyerabend P (1979) Science in a free society. Routledge

Feyerabend P (1984) Wissenschaft als Kunst. Suhrkamp, Frankfurt

Foley J et al (1982) Fundamentals of interactive computer graphics. Addison-Wesley, Reading, MA

Foresta D (1991) Monde Multiples. Editions Bas FEMIS, Paris

Foster H (ed) (1996) The return of the real: the avant-garde at the end of the century. MIT Press, Cambridge, MA

Frey G (1994) Anthropologie der Künste. Verlag Karl Alber, Freiburg

Gablik S (1984) Has modernism failed? Thames and Hudson Ltd., London

Gell-Mann M (1994) The quark and the jaguar. Adventures in the simple and the complex. W. H. Freeman and Company, New York

Gerbel K, Weibel P (eds) (1993) Genetische Kunst – Künstliches Leben. Genetic art – artificial life. Ars Electronica 93. PVS Verleger, Vienna

Gibson W (1995) Neuromancer. Voyager, London

Goldberg RL (1988) Performance art: from futurism to the present. Harry N. Abrams, New York

Golitsyn GA, Petrov VM (19959 Information and creation. Integrating the "Two Cultures". Birkhäuser, Boston

Gombrich EHJ (1995) The story of art. 16th Edition. Phaidon/Chronicle Books, London

Goodman C (1987) Digital visions: computers and art. Harry N. Abrams, New York

Goodman C (1996) InfoArt CD Rom. Rutt Video Interactive, http://www.rvi.com

Gould SJ (1980) The panda's thumb. W.W. Norton, New York

Haken H (1983) Advanced synergetics. Springer, Berlin Heidelberg New York Tokyo

Hall D, Fifer SJ (eds) (1990) Illuminating video: an essential guide to video art. Aperture Foundation, New York

Hanhardt JG (ed) (1986) Video culture. A critical investigation. New York

Henrich D, Iser W (1993) Theorien der Kunst. Suhrkamp Verlag, Frankfurt

Heisenberg W (1985) Gesammelte Werke – Collected Works. Blum W (ed) Springer, Berlin Heidelberg New York Tokyo

Heisenberg W (1980) Wandlungen in den Grundlagen der Naturwissenschaft. S. Hirzel Verlag, Stuttgart

Heisenberg W (1962) Physiscs and philosophy. Harper and Row, New York

Heisenberg W (1983) Tradition in science. Seabury Press

Heisenberg W (1974) Across the frontiers. Harpercollins

Herken R (1994) The universal turing machine. A half-century survey. Springer, Wien New York

Hershman L (ed) (1997) Clicking in. Hot links to a digital culture. San Francisco

Holland JH (1995) Hidden order. How adaptation builds complexity. Addison-Wesley, Reading, MA

Hünnekens A (1997) Der Bewegte Betrachter. Theorien der interaktiven Medienkunst. Wienand Verlag, Cologne

Husserl E (1991) Ding und Raum. Hahnengress KH et al. (ed) Felix Meiner Verlag, Hamburg

Husserl E (1980) Edmund Husserl phenomenology and the foundations of the sciences (His Ideas Pertaining to a Pure Phenomenology and to a Phenomenological Philosophy 3). Martinus Nijhoff

Kaiser G et al (1993) Kultur und Technik im 21. Jahrhundert. Campus Verlag, Frankfurt/New York

Kant I (1986) Critique of judgment. Free Press

Kant I (1996) Critique of practical reason (Great Books in Philosophy). Abbott T. K. (translator). Prometheus Books

Kauffman SA (1993) The origins of order. Oxford University Press, New York

Kepes GE (1995) Language of vision. Dover Publishers

Kepes GE (1965) Structure in art and in science. George Braziller

Kepes GE (1956) The new landscape in art and science. Paul Theobald, Chicago

Kittler FA, Johnston J (1997) Systems networks 1900–2000: Literature, Media, Information (Critical Voices). Distributed Art Publisher (Dap)

Klotz H (1996) Perspektiven der Medienkunst. Media Art Perspectives. Edition ZKM: Cantz Verlag

Klüver B (1966) Nine evenings: theater and engineering. New York Foundation for the Performing Arts

Klüver B et al (1972) Pavilion – experiments in art and technology. E.P. Dutton, New York

Knowlton K (1972) Collaborations with artists – a programmers reflection. North Holland Publishing

Kranz S (1974) Science and technology in the art. A tour through the realm of science/art. Van Nostrand Reinhold, New York

Krueger M (1991) Artificial reality II. Addison-Wesley, Reading, MA

Kuhn TS (1996) The structure of scientific revolutions. The University of Chicago Press, Chicago

Langton C (1989) Artificial life: the proceedings of an interdisciplinary workshop on the synthesis and simulation of living systems. Volume 6. 1 Edition. Mass.: Addison-Wesley

Langton C, Shimohara K (1997) Artificial life V: proceedings of the fifth international workshop on the synthesis and simulation of living systems (Complex Adaptive Systems). MIT Press, Cambridge, MA

Latour B, Woolgar S (1986) Laboratory life: the construction of scientific facts. Princeton University Press

Laurel B (1991) Computers as theater. Addison-Wesely, Menlo Park, CA

Levy S (1992) Artificial life. Vintage Books, New York

Levy S (1984) Hackers. Heroes of the computer revolution. New York

Lévi-Strauss C (1980) Mythos und Bedeutung. Suhrkamp Verlag, Frankfurt

Lovejoy M (1996) Postmodern currents: art and artists in the age of electronic media. Prentice Hall, New Jersey

Lyotard JF (1984) The postmodern condition: a report on knowledge. University of Minnesota Press, Minneapolis

McLuhan M (1994) Understanding media: the extensions of man. MIT Press, Cambridge, MA

McLuhan M, Powers BR (1992) The global village: transformations in world life and media in the 21st century. (Communication and Society (New York, N.Y.).) Oxford University Press

Maes P (ed) (1991) Designing autonomous agents: theory and practice from biology to engineering and back (Special Issues of Robotics and Autonomous Systems). Bradford Books

Malina FJ (ed) (1974) Kinetic art: theory and practice. Dover Publications, New York

Magnenat-Thalmann N, Magnenat-Thalmann D (eds) (1996) Interactive Computer Animation. Prentice Hall, New Jersey

Mandelbrot B (1988) The fractal geometry of nature. W. H. Freeman, New York

Moles AA (1971) Information theory and esthetic perception. University of Illinois Press, Urbana

Moser MA et al (1997) Immersed in technology. Art and virtual environments. MIT Press, Cambridge, MA

Mühlmann H (1996) The nature of cultures: A blueprint for a theory of culture genetics. Robert Payne (translator). Springer, Wien New York

Müller-Funk W et al (1996) Inszenierte Imagination – Beiträge zu einer historischen Anthropologie der Medien. Springer, Wien New York

Murphy MP, O'Neill LAJ (1995) What is life? The next fifty years. Cambridge University Press, Cambridge

Negroponte N (1996) Being digital. Vintage Books (Random House), New York

Negroponte N (1970) The architecture machine. MIT Press, Cambridge, MA

Nelson TH (1974) Computer lib/dream machines. Redmond, Washington

Peitgen H-O, Jürgens H, Saupe D, (1992) Chaos and fractals. New frontiers of science. Springer, Berlin Heidelberg New York Tokyo

Penny S (ed) (1995) Critical issues in electronic media. State University of New York

Popper F (1975) Art, action and participation. New York University Press

Popper F (1993) Art of the electronic age. Thames and Hudson, New York

Postman N (1993) Technopoly: the surrender of culture to technology. Vintage Books, New York

Prigogine I, Stengers I (1984) Order out of chaos. Bentam Books, New York

Prince P (1989) A brief history of Siggraph Art Exhibitions. Brave New Worlds. In: Leonardo Journal. Special edition, pp 3–5

Prusinkiewicz P, Lindenmayer A (1990) The algorithmic beauty of plants. Springer, Berlin Heidelberg New York Tokyo

Randell B (1982) The origins of digital computers. Springer, Berlin Heidelberg New York Tokyo

Ratliff F et al (1985) Visual arts and sciences. Transactions of the American Philosophical Society. Vol 75. Part 6. Am Philosophical Society

Reichardt J (1968) Cybernetic serendipity, the computer and the arts. Studio International, London

Reichardt J (1971) Cybernetics, art and ideas. Studio Vista and Van Nostrand, London, New York

Rheingold H (1993) The virtual community. Addison-Wesely, Reading, MA

Rheingold H (1991) Virtual reality. Summit, New York

Roszak T (1994) The cult of information: a neo-luddite treatise on high tech, artificial intelligence, and the true art of thinking. University of California Press

Rossler OE. Das Flammenschwert. Benteli Verlag, Bern

Rötzer F (1995) Die Telepolis. Bollmann Verlag

Schopenhauer A (1995) The world as will and idea. Everyman, London

Schwarz HP (1997) Media art history: Media Museum of the ZKM Karlsruhe. Prestel Verlag, München New York

Snow CP (1993) The two cultures. Cambridge University Press

Sontag S (1978) On photography. Anchor Books, New York

Stern DG (1995) Wittgenstein on mind and language. Oxford University Press

Steward I (1989) Does god play dice. Penguine Books

Steward I (1989) Game, set, and math. Basil Blackwell, Oxford

Stiles K, Selz P (1996) Contemporary art – a source book of artists' writings. University of California Press

Taiuti L (1996) Arte e Media. Costa & Nolan, Genova

Tisdall C (ed) (1992) Art meets science and spirituality in a changing economy. University of Washington Press

Turkle S (1984) The second self: computers and the human spirit. Simon and Schuster, New York

Virilio P (1991) The aesthetics of disappearance. Philip Beitchman (translator). Semiotext

Virilio P (1994) The vision machine. Julie Rose (translator). Indiana University Press

Wallis B, Tucker M (eds) (1994) Art after modernism: rethinking representation (Documentary Sources in Contemporary Art). David R. Godine

Weibel P (ed) (1995) Media pavillion: art and architecture in the age of cyberspace. Springer, Wien New York

Weibel P (1990) Probleme der Künstlichen Intelligenz. Merve Verlag, Berlin

Weizenbaum J (1984) Computer power and human reason: from judgement to calculation. Penguin

Wiener N (1988) The human use of human beings: cybernetics and society. Da Capo Press

Wilson EO (1984) Biophilia. Harvard University Press, Cambridge, MA

Wilson EO (1978) On human nature. Harvard University Press, Cambridge, MA

Wilson EO et al (1973) Life on earth. Sinauer Associates, Sunderland, MA

Wittgenstein L (1980) Culture and value. The University Press of Chicago, Chicago

Wittgenstein L (1967) Wittgenstein lectures and conversations on aesthetic. University of California Press

Woodfield R (ed) (1996) The Essential Gombrich: selected writings on art and culture by Ernst Hans Josef Gombrich. Phaidon/Chronicle Books, London

Journals and Exhibition Catalogues:

Ars Electronica. 1980 –. Exhibition catalogues and proceedings. Linz

Ars Electronica Center – Museum of the Future. 1996. Catalogue. Leopoltseder H, Stocker G (eds) Linz

Art Diary International. Giancarlo Politi Editore, Milan

Artec. Exhibition catalogues. Nagoya City Art Museum, Nagoya

Artifices. 1990 –. Exhibition catalogue. Boissier JL (ed) Saint Denis

Beyond Stereography. 1997. Exhibition catalogue. Moriyama T (ed) Tokyo Metropolitan Museum of Photography

Biennale d'Art Contemporain de Lyon. 1995. Exhibition catalogue. Musée d'Art Contemporain de Lyon

Cyberarts. Journal Exploring Art and Technology. San Francisco

EAT Bibliography. 1980. Klüver B(ed) New York

Europäisches Medienkunstfestival. 1988 –. Exhibition catalogues. Osnabrück

ICC Concept Book. 1997. Exhibition catalogue and proceedings. InterCommunication Center. Tokyo: NTT Publishing

IDEA International Directory for Electronic Arts. Bureaud A (ed) Chaos Editions, Paris

Images du Futur. 1991 –. Exhibition catalogues. Cité des Arts at des Nouvelles Technologies de Montréal

Imagination. 1995. Exhibition catalogue. Moriyama T (ed) Tokyo Metropolitan Museum of Photography

InfoArt – 95 Kwangju Biennale. 1995. Exhibition catalogue. Hong-hee K, Goodman C (eds) Kwangju Biennale Foundation

Inter Act – Schlüsselwerke Interaktiver Kunst. 1997. Exhibition catalogue. Wilhelm Lehmbruck Museum Duisburg

Interactions. 1990. Exhibition catalogue. Rijksmuseum Twenthe. Enschede

Interactive Garden. 1993. Exhibition catalogue. Huhtamo E (ed) Otso Gallery. Finland

Interactive Media Festival. 1994–1995. Exhibition catalogues. Los Angeles

InterCommunication. A Journal for Exploring the Frontiers of Art and Technology. NTT Publishing Co., Japan

ISEA International Symposium for the Electronic Arts. 1990 –. Exhibition catalogue and proceedings

Kunstforum International. Köln

Lab 5 and 6. International Film, Video and Computer Art Exhibition. 1995–1997. Exhibition catalogues. Kluszczynski RW (ed) Centrum for Contemporary Art. Warsaw

La Biennale di Venezia. 1995. Exhibition catalogue. 46. Esposizione internazionale d'arte. Marsilio

Leonardo Journal. MIT Press, Boston

Machine Culture: Siggraph Visual Proceedings. 1993. Exhibition catalogue. Penny S (ed) ACM Siggraph, New York

Machine Media. 1996. Exhibition catalogue of Vasulka, Steina and Woody. San Francisco Museum of Modern Art

Mediale Hamburg 1993. Exhibition catalogue

Multimediale. 1990 –. Exhibition catalogues. Zentrum für Kunst und Medientechnologie Karlsruhe

Oltre – Il Villaggio Globale. Beyond the Global Village. 1995. Exhibition catalogue. Mattei MG (ed) Electa, Milano

Press Enter. 1995. Exhibition catalogue. Power Plant Gallery Toronto

Prix Ars Electronica 1987 –. Exhibition catalogue. Leopoldseder H (ed) Linz

The Interaction. 1995–1997. Exhibition catalogues. Sakane I (ed) International Academy of Media Arts and Sciences. Gifu Japan

The Leonardo Almanac: International Resources in Art, Science, and Technology. MIT Press

Sky Art. 1984. Exhibition catalogue. MIT Cambridge

Siggraph Visual Proceedings. 1985 –. ACM Siggraph, New York

Software. 1970. Exhibition catalogue. Jewish Museum New York

Technics and Creativity. 1971. Museum of Modern Art, New York

Video Spaces: Eight Installations. 1995. Exhibition catalogue. London B (ed) Museum of Modern Art, New York

Wonderland of Science Art. 1989. Exhibition catalogue. Sakane I (ed) Kanagawa

WRO Media Art Festival. 1995 –. Exhibition catalogues. Wroclaw

Home Pages:

AEC Home Page: http://www.aec.at

ATR Home Page: http://www.atr.co.jp

CAiiA Home Page: http://caiiamind.nsad.newport.ac.uk

Exploratorium Home Page: http://www.exploratorium.edu

GMD Home Page: http://www.gmd.de

IAMAS Home Page: http://www.iamas.ac.jp

ICC-NTT Home Page: http://www.ntticc.or.jp

INM Home Page: http://www.inm.de

Interval Home page: http://web.interval.com/about

Santa Fe Institute, New Mexico, USA: http://www.santafe.edu

Siggraph Home Page: http://www.siggraph.org

ZKM Home Page: http://www.zkm.de

Index

SpringerArt

Gerfried Stocker, Christine Schöpf (eds.)

FleshFactor – Informationsmaschine Mensch

Ars Electronica Festival 97

1997. Numerous figures. 447 pages.
Soft cover DM 79,–, öS 550,–
Text: German/English
ISBN 3-211-82997-0

Since almost two decades Ars Electronica as a Festival for "Art, Technology and Society" has established its position as an important international forum for the discussion of the social-cultural dimensions of the digital revolution. The catalogue published for this Festival is characterized by the interdisciplinary dialogue among scientists, philosophers and artists and gives an extensive insight into the actual tendencies and developments. "FleshFactor – Informationsmaschine Mensch", the title of Ars Electronica 97, makes it clear that this year's theme is the "Mensch", the human being.
In light of the latest findings, developments and achievements in the fields of genetic engineering, neuro-science and networked intelligence, the conceptual complex now under investigation will include the status of the individual in networked artificial systems, the human body as the ultimate original, and the strategies for orientation and interrelation of the diametric opposites, man and machine, in the reciprocal, necessary processes of adaption and assimilation.

Springer-Verlag and the Environment

We at Springer-Verlag firmly believe that an international science publisher has a special obligation to the environment, and our corporate policies consistently reflect this conviction.

We also expect our business partners – printers, paper mills, packaging manufacturers, etc. – to commit themselves to using environmentally friendly materials and production processes.

The paper in this book is made from no-chlorine pulp and is acid free, in conformance with international standards for paper permanency.